Lecture Notes in Mathematics

Subseries:
Mathematisches Institut der Universität und
Max-Planck-Institut für Mathematik, Bonn - vol. 16

Adviser:
Friedrich Hirzebruch

Alexey A. Panchishkin

Non-Archimedean L-Functions

of Siegel and Hilbert Modular Forms

Springer-Verlag
Berlin Heidelberg New York
London Paris Tokyo
Hong Kong Barcelona
Budapest

Author

Alexey A. Panchishkin
Moscow State University
Department of Mathematics
Moscow, 119899-USSR

Mathematics Subject Classification (1980): Primary: 11F, 11R, 11S
Secondary: 19K, 46F, 46G

ISBN 3-540-54137-3 Springer-Verlag Berlin Heidelberg New York
ISBN 0-387-54137-3 Springer-Verlag New York Berlin Heidelberg

Printing and binding: Druckhaus Beltz, Hemsbach/Bergstr.
2146/3140-543210 - Printed on acid-free paper

Non-Archimedean L-Functions of Siegel and Hilbert modular forms

Content

Introduction 1

Acknowledgement 8

Chapter 1. Non-Archimedean analytic functions, measures and distributions. 9

§1. p-adic numbers and the Tate field 9
§2. Continuous and analytic functions over a non-Archimedean field 12
§3. Distributions, measures, and the abstract Kummer congruences 16
§4. Iwasawa algebra and the non-Archimedean Mellin transform 22
§5. Complex valued distributions, associated with Euler products 28

Chapter 2. Siegel modular forms and the holomorphic projection operator 35

§1. Siegel modular forms and Hecke operators 37
§2. Theta series, Eisenstein series and the Rankin zeta function 51
§3. Formulas for Fourier coefficients of Siegel – Eisenstein series 58
§4. Holomorphic projection operator and the Maass differential operator 70

Chapter 3. Non-Archimedean standard zeta functions of Siegel modular forms 81

§1. Description of the non-Archimedean standard zeta functions 81
§2. Complex valued distributions associated with standard zeta functions of Siegel modular forms 87
§3. Algebraic properties of the special values of normalized distributions 92
§4. Integrality properties and congruences for the distributions 110

Chapter 4. Non-Archimedean convolutions of Hilbert modular forms 117

§0. Introduction 117
§1. Hilbert modular forms 120
§2. Description of the non-Archimedean Rankin convolution of Hilbert automorphic forms 125
§3. Distributions on the Galois group Gal_S 126
§4. The integral representation of Rankin – Shimura and the holomorphic projection operator 133
§5. Integrality properties and congruences for the distributions 141

References 146

Subject Index 155

PREFACE

Main subject of the book is the arithmetic of zeta functions of automorphic forms. More precisely, we study p-adic properties of the special values of these functions. For the Riemann zeta function this goes back to the classical Kummer congruences for Bernoulli numbers and their p-adic interpretation given by Kubota – Leopoldt and Mazur. Using the technique of p-adic integration and the Rankin convolution method we construct the p-adic analytic continuation of the standard zeta functions of Siegel modular forms and of the convolutions of Hilbert modular forms. The book is intended for specialists in representation theory, functional analysis, algebraic geometry; it contains, together with new results, much background information about p-adic measures, their Mellin transforms, Siegel and Hilbert modular forms, Hecke operators, acting on them, Euler products etc. It seems that the method developed in the book is rather general and may have a number of further applications.

A.A.Panchishkin, Moscow, May 1991.

Introduction

The starting point in the theory of zeta functions is the expansion of the Rieman zeta-function $\zeta(s)$ into the Euler product:

$$\zeta(s) = \prod_p (1-p^{-s})^{-1} = \sum_{n=1}^{\infty} n^{-s} \quad (\mathrm{Re}(s) > 1)$$

The set of arguments s for which $\zeta(s)$ is defined can be extended to all $s \in \mathbf{C}, s \neq 1$, and we may regard $\mathbf{C}$ as the group of all continuous quasicharacters

$$\mathbf{C} = \mathrm{Hom}(\mathbf{R}_+^\times, \mathbf{C}^\times), \quad y \mapsto y^s$$

of $\mathbf{R}_+^\times$. The special values $\zeta(1-k)$ at negative integers are rational numbers:

$$\zeta(1-k) = -\frac{B_k}{k},$$

where B_k are Bernoulli numbers, which are defined by the formal power series equality

$$e^{Bt} = \sum_{n=0}^{\infty} \frac{B_n t^n}{n!} = \frac{t\, e^t}{e^t - 1},$$

and we know (by Sylvester – Lipschitz theorem, see [Mi-St], [Ka3]) that

$$c \in \mathbf{Z} \text{ implies } c^k(c^k-1)\frac{B_k}{k} \in \mathbf{Z}.$$

The theory of non-Archimedean zeta-functions originates in the work of Kubota and Leopoldt [Ku-Le] containing p-adic interpolation of these special values. Their construction turns out to be equivalent to classical Kummer congruences for the Bernoulli numbers, which we recall here in the following form (see [Ka3]). Let p be a fixed prime number, $c > 1$ an integer prime to p. Put

$$\zeta_{(p)}^{(c)}(-k) = (1-p^k)(1-c^{k+1})\zeta(-k)$$

and let $h(x) = \sum_{i=0}^{n} \alpha_i\, x^i \in \mathbf{Z}_p[x]$ be a polynomial over the ring $\mathbf{Z}_p$ of p-adic integers such that

$$x \in \mathbf{Z}_p \Longrightarrow h(x) \in p^m \mathbf{Z}_p.$$

Then we have that

$$\sum_{i=0}^{n} \alpha_i\, \zeta_{(p)}^{(c)}(-k) \in p^m \mathbf{Z}_p.$$

This property expresses the fact that the numbers $\zeta_{(p)}^{(c)}(-k)$ depend continuously on k in the p-adic sense; it can be deduced from the known formula for the sum of k-th powers:

$$S_k(N) = \sum_{n=1}^{N-1} n^k = \frac{1}{k+1}[B_{k+1}(N) - B_{k+1}]$$

in which $B_k(x) = (x+B)^k = \sum_{i=0}^{k} \binom{k}{i} B_i x^{k-i}$ denotes the Bernoulli polynomial. Indeed, all summands in $S_k(N)$ depend p-adic analytically on k, if we restrict ourselves to numbers n, prime to p, so that the desired congruence follows if we express the numbers $\zeta_{(p)}^{(c)}(-k)$ in terms of Bernoulli numbers (see [Ka3]).

The domain of definition of p-adic zeta functions is the p-adic analytic Lie group

$$X_p = \mathrm{Hom}_{\mathrm{contin}}(\mathbf{Z}_p^{\times}, \mathbf{C}_p^{\times})$$

of all continuous p-adic characters of the profinite group $\mathbf{Z}_p^{\times}$, where $\mathbf{C}_p = \widehat{\overline{\mathbf{Q}}}_p$ denotes the Tate field (completion of an algebraic closure of the p-adic field $\mathbf{Q}_p$), so that all integers k can be regarded as the characters $x_p^k : y \mapsto y^k$. The construction of Kubota and Leopoldt is equivalent to existence a p-adic analytic function $\zeta_p : X_p \to \mathbf{C}_p$ with a single pole at the point $x = x_p^{-1}$, which becomes a bounded holomorphic function on X_p after multiplication by the elementary factor $(x_p x - 1)$ $(x \in X_p)$, and is *uniquely determined* by the condition

$$\zeta_p({x_p}^k) = (1-p^k)\zeta(-k) \qquad (k \geq 1).$$

This result has a very natural interpretation in framework of the theory of non-Archimedean integration (due to Mazur): there exists a p-adic measure $\mu^{(c)}$ on $\mathbf{Z}_p^{\times}$ with values in $\mathbf{Z}_p$ such that $\int_{\mathbf{Z}_p^{\times}} x_p^k \, \mu^{(c)} = \zeta_{(p)}^{(c)}(-k)$. Indeed, if we integrate $h(x)$ over $\mathbf{Z}_p^{\times}$ we exactly get the above congruence. On the other hand, in order to define a measure $\mu^{(c)}$ satisfying the above condition it suffices for any continuous function $\phi : \mathbf{Z}_p^{\times} \to \mathbf{Z}_p$ to define its integral $\int_{\mathbf{Z}_p^{\times}} \phi(x) \, \mu^{(c)}$. For this purpose we approximate $\phi(x)$ by a polynomial (for which the integral is already defined), and then pass to the limit.

The important feature of the construction is that it equally works for primitive Dirichlet characters χ modulo a power of p: if we fix an embedding $i_p : \overline{\mathbf{Q}} \hookrightarrow \mathbf{C}_p$ then the character $\chi : (\mathbf{Z}/\mathbf{Z}_{p^N})^{\times} \to (\overline{\mathbf{Q}})^{\times}$ becomes an element of the torsion subgroup $X_p^{\mathrm{tors}} \subset X_p$ and the above equality also holds for the special values $L(-k, \chi)$ of the Dirichlet L-series

$$L(s, \chi) = \sum_{n=1}^{\infty} \chi(n) \, n^{-s} = \prod_p (1 - \chi(p)p^{-s})^{-1},$$

so that we have

$$\zeta_p(\chi x_p^k) = i_p[(1-\chi(p)p^k)L(-k, \chi)] \quad (k \geq 1, \ k \in \mathbf{Z}, \ \chi \in X_p^{\mathrm{tors}}).$$

The original construction of Kubota and Leopoldt [Ku-Le] was successesfully used by Iwasawa [Iw] for the description of the class groups of cyclotomic fields. Since then the class of functions admitting p-adic analogues has gradually extended.

Zeta-functions (of complex variable) can be attached as certain Euler products to various objects such as diophantine equations, representations of Galois groups, modular forms etc., and they play a crucial role in modern number theory. Deep interrelations between these objects discovered in last decades are based on identities for the corresponding zeta functions which presumably all fit into a general concept of the Langlands of L-functions associated with automorphic representations of a reductive group G over a number field K. From this point of view the study of arithmetic properties of these zeta functions is becoming especially important.

The theory of modular symbols (due to Mazur and Manin, see [Man1]–[Man6], [Maz-SD]) provided a non-Archimedean construction of functions, which correspond to the case of the group $G = \mathrm{GL}_2$ over $K = \mathbf{Q}$. Several authors (including Deligne, Ribet, N.M.Katz, Kurčanov and others, see [De-Ri], [Ka1]–[Ka3], [Kurč1] –[Kurč3], [Sho], [V1], [V2]) investigated this problem for the case $G = \mathrm{GL}_1$ and GL_2 over totally real fields and fields of CM-type (i.e. totally imaginary quadratic extentions of totally real fields). But the case of more general reductive groups remained unclear until the mid-eighties although important complex analytic properties of the Langlands L-functions had been proved. In recent years a general approach to construction of non-Archimedean L-functions associated with various classes of automorphic forms was developed, in particular, for the case of symplectic groups of even degree over $K = \mathbf{Q}$ and the group $G = \mathrm{GL}_2 \times \mathrm{GL}_2$ over a totally real field K.

The main tool of the appearing theory is the systematic use of the Rankin-Selberg method for obtaining both complex-valued and p-adic distributions as certain integrals involving cusp forms and Eisenstein series. By this method we constructed non-Archimedean analogues of the standard zeta functions attached to Siegel cusp forms of even degree and of sufficiently large weight.

For a Siegel modular form $f(z)$ of degree m and weight k, which is an eigenfunction of the Hecke algebra, and for each prime number p one can define the Satake p-parameters of f denoted by $\alpha_i(p)$ with $i = 0, 1, \cdots, m$. In this introduction we assume for simplicity that f is a modular form with respect to the whole Siegel modular group $\Gamma^m = \mathrm{Sp}_m(\mathbf{Z})$, but in chapter 3 we treat the greneral case of forms of level C with a Dirichlet character ψ mod C. The standard zeta function of f is defined by means of the Satake p-parameters as the following Euler product:

$$\mathcal{D}(s, f, \chi) =$$
$$= \prod_p \left\{ \left(1 - \frac{\chi(p)}{p^s}\right) \prod_{i=1}^m \left(1 - \frac{\chi(p)\alpha_i(p)}{p^s}\right) \left(1 - \frac{\chi(p)\alpha_i(p)^{-1}}{p^s}\right) \right\}^{-1},$$

where χ is an arbitrary Dirichlet character. According to A.N.Andrianov and V.L.Kalinin [An-K], this function can be represented in a form of an integral convolution of f and a theta series with a Siegel – Eisenstein series as a kernel. The construction of its p-adic analytic continuation is based on explicit formulas for the special values of the standard zeta function, and this result is equivalent to the existence of some generalized Kummer congruences for these values. Before giving the precise statements of our results we introduce the normalized zeta functions

$$\mathcal{D}^-(s, f, \chi) = (2\pi)^{-m(s+k-(m+1)/2)} \prod_{j=1}^m \Gamma(s + k - j) \mathcal{D}(s, f, \chi),$$

$$\mathcal{D}^+(s, f, \chi) = \frac{2i^\delta \Gamma(s) \cos(\pi(s - \delta)/2)}{(2\pi)^s} \mathcal{D}^-(s, f, \chi),$$

$$\mathcal{D}^*(s, f, \chi) = \pi^{-(s+\delta)/2} \Gamma((s + \delta)/2) \mathcal{D}^-(s, f, \chi),$$

where $\delta = 0$ or 1 according as $\chi(-1) = 1$ or $\chi(-1) = -1$, and let

$$f(z) = \sum_{\xi>0} a(\xi)e_m(\xi z) \in \mathcal{S}_k^m$$

be the Fourier expansion of the Siegel cusp form $f(z)$ of weight k, the sum is extended over all positive definite half integral $m \times m$-matrices, $z \in \mathfrak{H}_m$,

$$\mathfrak{H}_m = \{z \in \mathrm{GL}_m(\mathbf{C}) \mid {}^t z = z, \;\; \mathrm{Im}(z) \text{ is positive definite}\}$$

be the Siegel upper half plane of degree m and $e_m(z) = \exp(\mathrm{tr}(2\pi i z))$. Assume that $k > 2m+2$ and m is even.

Theorem A *(Algebraic properties of the special values of the standard zeta functions)*

a) For all integers s with $1 \le s \le k-\delta-m$ and Dirichlet characters χ such that χ^2 is non-trivial for $s=1$ we have that

$$\langle f, f\rangle^{-1}\mathcal{D}^+(s, f, \chi) \in K = \mathbf{Q}(f, \Lambda_f, \chi),$$

where $K = \mathbf{Q}(f, \Lambda_f, \chi)$ denotes the field generated by Fourier coefficients of f, by the eigenvalues $\Lambda_f(X)$ of Hecke operators X on f, and by the values of the character χ.

b) For all integers s with $1-k+\delta+m \le s \le 0$ we have that

$$\langle f, f\rangle^{-1}\mathcal{D}^-(s, f, \chi) \in K.$$

We may suppose without loss of generality that $a(\xi) \in \overline{\mathbf{Q}}$ for all $\xi > 0$, and let us assume also that $a(\xi_0) = 1$ for some $\xi_0 > 0$ with $\det 2\xi_0 = 1$. Moreover, we make the essential assumption that the form f is *p-ordinary* in the sense that $\mid i_p(\alpha_0(p) \mid_p = 1$ for a fixed embedding $i_p : \overline{\mathbf{Q}} \hookrightarrow \mathbf{C}_p$, and that $a_0(\xi_0) \neq 0$ for the Fourier coefficient $a_0(\xi_0)$ of the p-auxiliary form

$$f_0(z) = \sum_{\xi>0} a_0(\xi)e_m(\xi z)$$

defined in §1 of chapter 2 (see (1.47)). According to its definition, this form satisfies the following multiplicativity property

$$a_0(p^N\xi) = \alpha_0(p)^N a_0(\xi).$$

The importance of the above non-vanishing condition of $a_0(\xi_0)$ was pointed out to the author by S. Böcherer.

Theorem B *(non-Archimedean interpolation of the standard zeta functions) Under the assumptions as above for each integer $c > 1$ prime to p there exist bounded $\mathbf{C}_p$-analytic functions*

$$\mathcal{D}^{c+}(x, f), \;\; \mathcal{D}^{c-}(x, f) : X_p \to \mathbf{C}_p,$$

which are uniquely determined by the following conditions:

a) for all non-trivial Dirichlet characters $\chi \in X_p^{\mathrm{tors}}$ and for all integers s with $1 \le s \le k - \delta - m$ the following equality holds

$$\mathcal{D}^{c+}(\chi x_p^s, f) =$$
$$i_p \left(\frac{G_m(\chi)\, C_\chi^{m(s+k-1-m)}}{\alpha_0(C_\chi)^2} \frac{C_\chi^s}{G(\bar\chi)} (1 - \bar\chi^2(c) c^{-2s}) \frac{\mathcal{D}^+(s, f, \bar\chi)}{\langle f,\ f\rangle} \right),$$

b) for all non-trivial Dirichlet characters $\chi \in X_p^{\mathrm{tors}}$ and for all integers s with $1 - k + \delta + m \le s \le 0$ the following equality holds

$$\mathcal{D}^{c-}(\chi x_p^s, f) =$$
$$i_p \left(\frac{G_m(\chi)\, C_\chi^{m(s+k-1-m)}}{\alpha_0(C_\chi)^2} (1 - \chi^2(c) c^{2s-2}) \frac{\mathcal{D}^-(s, f, \bar\chi)}{\langle f,\ f\rangle} \right),$$

where

$$G_m(\chi) = \sum_{h \in M_m(\mathbf{Z}) \bmod C_\chi} \chi(\det h) e_m(h/C_\chi)$$

denotes the Gauss sum of degree m of the primitive Dirichlet character $\chi \bmod C_\chi$, $C_\chi = p^{N_\chi}$, $\alpha_0(C_\chi) = \alpha_0(p)^{N_\chi}$, $G(\chi) = G_1(\chi)$.

The standard zeta function $D(s, f, \chi)$ provides an example for the general definition of the Langlands L-functions. For a reductive group G over a number field K this definition is based on the notion of the Langlands L–group ${}^LG$; this group is a complex analytic reductive group such that the lattice of characters of its maximal torus and the lattice of its one-parameter subgroups (cocharacters) are obtained from the analogous objects of the group G by inversion.

The important fact in the representation theory of reductive groups over local fields is that semisimple conjugacy classes h_v of ${}^LG$ for a place v of K correspond to certain infinite dimensional representations π_v of the group $G(K_v)$ over the local field K_v (the completion of K at v). It is known that for groups of the type $\mathbf{A}_n$ and $\mathbf{D}_n$ this construction preserves their types, and interchanges the types $\mathbf{B}_n$ and $\mathbf{C}_n$, so that if $G = \mathrm{GL}_n$ then ${}^LG = \mathrm{GL}_n(\mathbf{C})$, and if $G = \mathrm{GSp}_m$ then ${}^LG = \mathrm{Spin}_{2m+1}(\mathbf{C})$, the universal covering of the orthogonal group $\mathrm{SO}_{2m+1}(\mathbf{C})$. For example, if $G = \mathrm{GL}_2$ and v a non-Archimedean place then ${}^LG = \mathrm{GL}_2(\mathbf{C})$ and for $h_v \ni \begin{pmatrix} \alpha_1 & 0 \\ 0 & \alpha_2 \end{pmatrix}$ the corresponding representation π_v is the representation $\mathrm{Ind}_T^G(\mu_1 \otimes \mu_2)$, which is induced from the maximal torus $T = \mathrm{GL}_1 \times \mathrm{GL}_1$, μ_1, $\mu_2 : K_v^\times \to \mathbf{C}^\times$ being unramified quasicharacters of $K_v^\times$ with $\mu_i(\mathfrak{p}_v) = \alpha_i$, $i = 1, 2$.

Let π be an automorphic representation of the group G, which is an irreducible subrepresentation of the smooth regular representation of the adelic group $G(\mathbf{A}_K)$. Then there is the decomposition of π into the infinite tensor product: $\pi = \otimes_v \pi_v$ where π_v is a representation of $G(K_v)$ which correspond to certain classes h_v from ${}^LG$ for almost all v (i.e for $v \notin S$ where S is a finite set of places of K). For a finite dimensional representation $r: {}^LG \to \mathrm{GL}_t(\mathbf{C})$ of the L-group we define automorphic L-functions

$$L(s,\pi,r) = L_S(s,\pi,r) = \prod_{v\notin S} \det(1_t - (\mathcal{N}v)^{-s} r(h_v))^{-1}$$

where $\mathcal{N}v$ is the number of elements of the residue class field of v (which is a power of its characteristic), and the product is taken over all non-Archimedean places v, $v \notin S$.

In the Siegel modular case we consider, associated with f, the automorphic representation π_f, which is generated by a function on $\mathrm{GSp}_m(\mathbf{A})$ inflated from the cusp form f on $\mathfrak{H}_m$ (as a subrepresentation of the regular representation of $G(\mathbf{A_Q}) = \mathrm{GSp}_m(\mathbf{A})$). The irreducibility of π_f is equivalent to the fact that f is an eigenfunction for the Hecke algebra $\mathcal{H}^m = \otimes_p \mathcal{H}_p^m$ of the Siegel modular group Γ_m of degree m. In this case the corresponding character of $\mathcal{H}_p$ on f is completely determined by its Satake p-parameters, and for the universal covering $r: \mathrm{Spin}_{2m+1}(\mathbf{C}) \to \mathrm{SO}_{2m+1}(\mathbf{C})$ with $\mathrm{Spin}_{2m+1}(\mathbf{C}) \subset \mathrm{GL}_{2^m}(\mathbf{C})$ we have that the classes h_v and $r(h_v)$ are represented by the matrices

$$h_v = Sp(h_v) \ni \mathrm{diag}\{\alpha_0(p)\alpha_{i_1}(p)\cdots\alpha_{i_r}(p) \mid 0 \le r \le m,\ 1 \le i_1 < \ \cdots\ i_r \le m\}$$
$$r(h_v) = St(h_v) \ni \mathrm{diag}\{1, \alpha_1(p), \alpha_2(p), \cdots, \alpha_m(p), \alpha_1(p)^{-1}, \alpha_2(p)^{-1}, \cdots, \alpha_m(p)^{-1}\},$$

where Sp and St are called, respectively, *spinor* and *standard* representations of the Langlands group ${}^LG$. Therefore the standard zeta function $D(s,f,\chi)$ coincides with the L-function $L(s,\pi_f,St)$. The function

$$L(s,\pi_f,Sp) = \prod_p [\prod_{\substack{0\le r\le m\\ 1\le i_1< \cdots\ i_r\le m}} (1-\chi(p)\alpha_0(p)\alpha_{i_1}(p)\cdots\alpha_{i_r}(p)p^{-s})]^{-1}$$

is the *spinor zeta function* of f. Its analytic properties were studied by A. N. Andrianov in the case $m = 2$ but stil little is known about algebraic and arithmetic properties of the special values of this function; however, from the general Deligne conjecture on critical values of L-functions it follows that the properties analogous to those given in Theorem A could exist only for $s = k-1$.

In the last chapter we construct non-Archimedean convolutions of two Hilbert modular forms of different (scalar) weight. The exposition here is provided with some basic facts about Hilbert modular forms. Let p be a prime, and let S be a finite set of primes which contains p. Convolutions of Hilbert modular forms correspond to certain automorphic representationson the group $G = \mathrm{GL}_2 \times \mathrm{GL}_2$ over a totally real field F, and these convolutions reduce to zeta functions of the form

$$L(s,\mathbf{f},\mathbf{g}) = \sum_{\mathfrak{n}} C(\mathfrak{n},\mathbf{f})\, C(\mathfrak{n},\mathbf{g}) \mathcal{N}(\mathfrak{n})^{-s},$$

where $\mathbf{f}$, $\mathbf{g}$ are *Hilbert automorphic forms of "holomorphic type"* over F, where $C(\mathfrak{n},\mathbf{f})$, $C(\mathfrak{n},\mathbf{g})$ are their normalized Fourier coefficients (indexed by integral ideals $\mathfrak{n}$ of the maximal order $\mathcal{O}_F \subset F$). We regard $\mathbf{f}$, $\mathbf{g}$ as functions on the adelic group $G_\mathbf{A} = \mathrm{GL}_2(\mathbf{A}_F)$, where $\mathbf{A}_F$ is the ring of adeles of F and we suppose that $\mathbf{f}$ is a *primitive cusp form* of scalar weight $k \geq 2$, of conductor $\mathfrak{c}(\mathbf{f}) \subset \mathcal{O}_F$, and the character ψ and $\mathbf{g}$ a primitive cusp form of weight $l < k$, the conductor $\mathfrak{c}(\mathbf{g})$, and the character ω, $(\psi, \omega : \mathbf{A}_F^\times \to \mathbf{C}^\times$ are Hecke characters of finite order). The non-Archimedean construction is based on the algebraic properties of the special values of the function $L(s, \mathbf{f}, \mathbf{g})$ at the points $s = l, \cdots, k-1$ up to some constant, which is expressed in terms of the Petersson inner product $\langle \mathbf{f}, \ \mathbf{f} \rangle$ of the automorphic form $\mathbf{f}$ [Shi6]. Our theorem on non-Archimedean interpolation is equivalent to certain generalized Kummer congruences for these special values.

We need some more notation for the precise formulation of the result (in a simplified form). Let ψ^*, ω^* be the characters of the ideal group of F associated with ψ, ω and let

$$L_{\mathfrak{c}}(s, \psi\omega) = \sum_{\mathfrak{n}+\mathfrak{c}=\mathcal{O}_F} \psi^*(\mathfrak{n})\omega^*(\mathfrak{n})\mathcal{N}(\mathfrak{n})^{-s} = \prod_{\mathfrak{p}+\mathfrak{c}=\mathcal{O}_F} (1 - \psi^*(\mathfrak{p})\omega^*(\mathfrak{p})\mathcal{N}(\mathfrak{p})^{-s})^{-1}$$

be the corresponding Hecke L-function with $\mathfrak{c} = \mathfrak{c}(\mathbf{f})\mathfrak{c}(\mathbf{g})$. We now define the normalized Rankin zeta function by setting

$$\Psi(s, \mathbf{f}, \mathbf{g}) = \gamma_n(s) L_{\mathfrak{c}}(2s + 2 - k - l, \psi\omega) L(s, \mathbf{f}, \mathbf{g}),$$

where $n = [F : \mathbf{Q}]$ is the degree of F,

$$\gamma_n(s) = (2\pi)^{-2ns}\Gamma(s)^n\Gamma(s + 1 - l)^n$$

is the gamma-factor. Then the function $\Psi(s, \mathbf{f}, \mathbf{g})$ admits a holomorphic analytic continuation onto the entire comlex plane and it satisfies a certain functional equation [Ja]. Put $\Omega(\mathbf{f}) = \langle \mathbf{f}, \ \mathbf{f} \rangle_{\mathfrak{c}(\mathbf{f})}$, then we know due to Shimura [Sh6] that the number

$$\frac{\Psi(l + r, \mathbf{f}, \mathbf{g})}{(2\pi i)^{n(1-l)}\Omega(\mathbf{f})} \text{ is algebraic for all integers } r \text{ with } 0 \leq r \leq k - l - 1$$

For the non-Archimedean construction we introduce the S-adic completion

$$\mathcal{O}_S = \prod_{q \in S} (\mathcal{O}_F \otimes \mathbf{Z}_q) = \prod_{\mathfrak{p} | q \in S} \mathcal{O}_{\mathfrak{p}}$$

of the ring $\mathcal{O}_F$. Put

$$S_F = \{\mathfrak{p} \mid \mathfrak{p} \ \text{ divides } \ q \in S\}$$

and let $\mathrm{Gal}_S = \mathrm{Gal}(F(S)/F)$ be the Galois group of the maximal abelian extension of F unramified outside S and ∞.

The domain of definition of our non-Archimedean L-functions is the p-adic analytic Lie group

$$\mathcal{X}_S = \mathrm{Hom}_{\mathrm{contin}}(\mathrm{Gal}_S, \mathbf{C}_p^\times)$$

of all continuous p-adic characters of the Galois group Gal_S with $\mathbf{C}_p$ being the Tate field. Elements of finite order $\chi \in \mathcal{X}_S$ can be identified with those Hecke characters of finite order whose conductors are divisible only by prime divisors belonging to S_F, via the decomposition

$$\chi : \mathbf{A}_F^\times \stackrel{\text{class field theory}}{\longrightarrow} \mathrm{Gal}_S \to \overline{\mathbf{Q}}^\times \stackrel{i_p}{\to} \mathbf{C}_p^\times.$$

We shall use the same symbol χ to denote both Hecke character and the corresponding element of $\mathcal{X}_S$. Since $\mathbf{Q}(S) \subset F(S)$, the restriction of Galois automorphisms to $\mathbf{Q}(S)$ determines a natural homomorphism

$$\mathcal{N} : \mathrm{Gal}_S \to \mathrm{Gal}(\mathbf{Q}(S)/\mathbf{Q}) \cong \mathbf{Z}_S^\times = \prod_{q \in S} \mathbf{Z}_q^\times .$$

We shall let $\mathcal{N} x_p$ denote the composition of this homomorphism with the natural projection $\prod_q \mathbf{Z}_q^\times \to \mathbf{Z}_p^\times$ and the inclusion $\mathbf{Z}_p^\times \subset \mathbf{C}_p^\times$.

Again our essential assumption is that that the cusp form $\mathbf{f}$ is *p-ordinary*, i.e. for the fixed embedding $\overline{\mathbf{Q}} \hookrightarrow \mathbf{C}_p$ and for all $\mathfrak{p} \mid p$ there exists such a root $\alpha(\mathfrak{p})$ of Hecke $\mathfrak{p}$-polynomial of $\mathbf{f}$ that $| i_p(\alpha(\mathfrak{p})) |_p = 1$. We then fix these roots $\alpha(\mathfrak{p})$ and extend the definition of $\alpha(\mathfrak{m})$ to all integral ideals $\mathfrak{m} \subset \mathcal{O}_F$ by multiplicativity.

Theorem C *(On the non-Archimedean convolutions of Hilbert modular forms) Under the above notations and assumptions there exists a bounded $\mathbf{C}_p$-analytic function $\Psi_S : \mathcal{X}_S \to \mathbf{C}_p$, which is uniquely determined by the condition: for each Hecke character of finite order $\chi \in \mathcal{X}_S^{\mathrm{tors}}$ the following equality holds:*

$$\Psi_S(\chi) = i_p \left[D_F^{2l} \omega(\mathfrak{m}) \frac{\tau(\chi)^2 \mathcal{N} \mathfrak{m}^{l-1}}{\alpha(\mathfrak{m})^2} \frac{\Psi(l, \mathbf{f}, \mathbf{g}^\rho(\bar{\chi}))}{(-2\pi i)^{n(1-l)} \langle \mathbf{f}, \mathbf{f} \rangle} \right],$$

where D_F is the discriminant of F, $\tau(\chi)$ is the Gauss sum of χ, and $\mathbf{g}^\rho(\chi)$ the cusp form obtained from $\mathbf{g}$ by complex conjugation of its Fourier coefficients and by twisting it then with the character χ.

The analogous result is also valid for the special values $\Psi(l+r, \mathbf{f}, \mathbf{g})$ with $r = 1, \ldots, k-l-1$, if we replace $\chi \in \mathcal{X}_S$ by $\chi \mathcal{N} x_p^r \in \mathcal{X}_S$ (see the Main theorem in §2 of chapter 4).

In the case of the ground field $F = \mathbf{Q}$ this result was established by H.Hida [Hi1] and by the author [Pa9], [Pa10].

Recently this construction was extended by My Vinh Quang (Moscow University) to the *non-p-ordinary*, i.e. *supersingular* case, when $| i_p(\alpha(\mathfrak{p})) |_p < 1$ for both roots $\alpha(\mathfrak{p})$ and for all $\mathfrak{p} \mid p$. In this situation the function Ψ_S is also uniquely determined by the above condition provided that it has only a prescribed logarithmic growth on $\mathcal{X}_S$.

Acknowledgement. I would like to express my deep gratitude to Professor F.Hirzebruch and to the Max – Planck – Institut für Mathematik in Bonn for the hospitality and for the support during my visits to the MPI in summer 1989 and in spring 1990. It is a pleasure for the author to thank Professors H.Klingen, S.Böcherer, J.Elstrodt, G.Harder, W.Kohnen, P.Schneider, N.- P.Skoruppa and Don B. Zagier for very helpfull discussions and advice. Also I am grateful to Prof. D.Barsky , Prof. A.Klopotowski and to the University Paris – Nord for the help in making final corrections to the manuscript.

Chapter 1. Non-Archimedean analytic functions, measures and distributions.

In this chapter we give an exposition of some standard facts from the theory of continuous and analytic functions over a non-Archimedean local field. We start by recalling the definitions and notations concerning p-adic and S-adic numbers. Then we discuss the theory of continuous p-adic functions and their p-adic interpolation, and also the basic properties of p-adic analytic functions. In §3 we introduce distributions and measures and give a general criterion for the existence of a non-Archimedean measure with given values of integrals of functions belonging to certain dense family ("generalized Kummer congruences"). The next §4 is devoted to a description of the algebra of bounded measures in terms of their non-Archimedean Mellin transforms (Iwasawa isomorphism). The chapter is completed with an exposition of a general construction of measures, attached to rather arbitrary Euler products.This construction provides a generalization of measures first introduced by Yu.I.Manin [Man4], B.Mazur and H.P.F.Swinnerton–Dyer [Maz-SD]. Our construction [Pa5], [Pa9] was already successfully used in several problems concerning the p-adic analytic interpolation of special values of Dirichlet series [Ar], [Co-Schm], [Co-Schn], [Sch].

§1.p-adic numbers and the Tate field

1.1. Let p be a prime number, $\mathbf{Q}_p$ the field of p-adic numbers, i.e. the completion of the field of rational numbers $\mathbf{Q}$ with respect to the p-adic metric, given by the p-adic valuation

$$|\cdot|_p : \mathbf{Q} \to \mathbf{R}_{\geq 0} = \{x \in \mathbf{R} \mid x \geq 0\}$$
$$|a/b|_p = p^{\mathrm{ord}_p b - \mathrm{ord}_p a}, \quad |0|_p = 0,$$

where $\mathrm{ord}_p a$ is the highest power of p dividing the integer a. The function $|\cdot|_p$ is multiplicative, since

$$\mathrm{ord}_p(xy) = \mathrm{ord}_p x + \mathrm{ord}_p y, \tag{1.1}$$

and satisfies the non-Archimedean property

$$|x+y|_p \leq \max(|x|_p, |y|_p). \tag{1.2}$$

If K is a finite algebraic extension of $\mathbf{Q}_p$ then K is generated over $\mathbf{Q}_p$ by a primitive element $\alpha \in K$, so that α is a root of an irreducible polynomial of degree $d = [K : \mathbf{Q}_p]$,

$$f(x) = x^d + a_{d-1}x^{d-1} + \ldots + a_0 \in \mathbf{Q}_p[x].$$

The valuation $|\cdot|_p$ admits a unique extention to K defined by

$$|\beta|_p = (|\mathcal{N}_{K/\mathbf{Q}_p}(\beta)|_p)^{1/d}, \tag{1.3}$$

where $\mathcal{N}_{K/\mathbf{Q}_p}(\beta) \in \mathbf{Q}_p$ is the algebraic norm of an element $\beta \in K$. The formula (1.3) defines a unique extension of $|\cdot|_p$ to the algebraic closure $\overline{\mathbf{Q}}_p$ of $\mathbf{Q}_p$, which satisfies (1.2) (see [Kol]). The function ord_p is then also extended to $\overline{\mathbf{Q}}_p$ by $\mathrm{ord}_p\alpha = \log_p |\alpha|_p$. The formula (1.3) implies that $\mathrm{ord}_p K^\times$ is an additive subgroup in $\frac{1}{d}\mathbf{Z}$, hence $\mathrm{ord}_p K^\times = \frac{1}{e}\mathbf{Z}$ for a certain positive integer e dividing d which is called *the ramification index* of the extension $K/\mathbf{Q}_p$.

Put

$$\mathcal{O}_K = \{x \in K \,|\, |x|_p \le 1\}\ ,\ M_K = \{x \in K \,|\, |x|_p < 1\} \tag{1.4}$$

Then M_K is the maximal ideal in $\mathcal{O}_K$ and the residue field $\mathcal{O}_K/M_K$ is a finite extension of degree f of $\mathbf{F}_p$ and there is the relation $d = e \cdot f$, in which f is called *the inertial degree* of the extension. For each $x \in \mathcal{O}_K$ its *Teichmüller representative* is defined by

$$\omega(x) = \lim_{n\to\infty} x^{p^{fn}},\quad \omega(x) \equiv x\,(\mathrm{mod}\, M_K), \tag{1.5}$$

and satisfies the equation

$$\omega(x)^{p^f} = \omega(x).$$

The map ω provides a homomorphism of the group of invertible elements

$$\mathcal{O}_K^\times = \mathcal{O}_K/M_K = \{x \in K \,|\, |x|_p = 1\}$$

of $\mathcal{O}_K$ onto the group of roots of unity of degree $p^f - 1$ in K, denoted by μ_{p^f-1}, and the isomorphism

$$(O_K/M_K)^\times \xrightarrow{\sim} \mu_{p^f-1} \subset \mathcal{O}_K^\times. \tag{1.6}$$

For example, if $e = 1$ then the extension K is called *unramified*. In that case $f = d$ and the Teichmüller representatives generate K over $\mathbf{Q}_p$, therefore

$$K = \mathbf{Q}_p(1^{1/N}),\quad N = p^d - 1.$$

On the other hand, if $e = d$ then the extension K is called *totally ramified*. For example, if ζ is a primitive root of unity of degree p^n, then $\mathbf{Q}_p(\zeta)$ is totally ramified of degree $d = p^n - p^{n-1}$, and we have that

$$\mathrm{ord}_p(\zeta - 1) = \frac{1}{p^n - p^{n-1}}. \tag{1.7}$$

1.2. For a field K with the non-Archimedean valuation $|\cdot|_p$ let us define

$$D_a(r) = D_a(r;K) = \{x \in K \,|\, |x-a|_p \le r\}, \tag{1.8}$$

$$D_a(r^-) = D_a(r^-;K) = \{x \in K \,|\, |x-a|_p < r\} \tag{1.9}$$

("closed" and "open" discs of the radius r with the center at the point $a \in K$, $r \ge 0$). Then we have that $D_b(r) = D_a(r)$ for any $b \in D_a(r)$ and $D_b(r^-) = D_a(r^-)$ for any $b \in D_a(r^-)$. Note that in the topological sense both discs (1.8) and (1.9) are open and closed subsets of the topological field K.

The important property of the field $\mathbf{Q}_p$ and of its finite extensions is that they are locally compact, so that the discs (1.8) and (1.9) are compact. The disc

$$\mathbf{Z}_p = D_0(1;\mathbf{Q}_p) = \{x \in \mathbf{Q}_p \,|\, |x| \le 1\}$$

is a compact topological ring (the ring of p-adic integers), which is isomorphic to the projective limit of residue rings $\mathbf{Z}/p^n\mathbf{Z}_p$ with respect to the homomorphisms of reduction modulo p^n:

$$\mathbf{Z}_p = \varprojlim_n \mathbf{Z}/p^n\mathbf{Z}_p,$$

Analogously, we have that

$$\mathcal{O}_K = \varprojlim_n \mathcal{O}_K/M_K^n,$$

and

$$\mathbf{Z}_p^\times = \varprojlim_n (\mathbf{Z}/p^n\mathbf{Z}_p)^\times, \quad \mathcal{O}_K^\times = \varprojlim_n (\mathcal{O}_K/M_K^n)^\times$$

(the projective limit of groups).

1.3. The structure of the multiplicative group $\mathbf{Q}_p^\times$ and $K^\times$. Put $\nu = 1$ for $p > 2$ and $\nu = 1$ for $p = 2$, and define

$$U = U_p = \{x \in |x \equiv 1 \bmod p^\nu\}. \tag{1.10}$$

Then there is an isomorphism $U \xrightarrow{\sim} \mathbf{Z}_p$ of the multiplicative group U_p and of the additive group $\mathbf{Z}_p$ which is provided by combining the natural homomorphism

$$U \xrightarrow{\sim} \varprojlim_n U/U^{p^n}$$

and by special isomorphisms

$$\alpha_{p^n} : U/U^{p^n} \xrightarrow{\sim} \mathbf{Z}/p^n\mathbf{Z},$$

given by

$$\alpha_{p^n}((1+p^\nu)^a) = a \bmod p^n \quad (a \in \mathbf{Z}). \tag{1.11}$$

One easily verifies that (1.11) is well defined and gives the desired isomorphism. Therefore, the group U is a topological cyclic group, and $1+p^\nu$ can be taken as its generator. There are the following decompositions

$$\mathbf{Q}_p = p^{\mathbf{Z}} \times \mathbf{Z}_p^\times, \quad \mathbf{Z}_p^\times \cong (\mathbf{Z}/p^\nu\mathbf{Z})^\times \times U. \tag{1.12}$$

Similarly, if $[K : \mathbf{Q}_p] = d$, then

$$K^\times = \pi^{\mathbf{Z}} \times \mathcal{O}_K^\times, \quad \mathcal{O}_K^\times \cong (\mathcal{O}_K/M_K)^\times \times U_K,$$

where π is a generator of the principial ideal $M_K = \pi\mathcal{O}_K$ (i.e. any element $\pi \in K^\times$ with $\mathrm{ord}_p\pi = 1/e$),

$$U_K = \{x \in \mathcal{O}_K^\times \,|\, |x-1|_p < 1\} = D_1(1^-; K).$$

The structure of the group U_K is then described as a direct product of d copies of the additive group $\mathbf{Z}_p$ and a finite group consisting of all p-primary roots of unity contained in K (see [Ca-F], [Kol]).

1.4. The S-adic numbers. Let S be a finite set of prime numbers.For a positive integer M denote by $S(M)$ its support, i.e. the set of all prime numbers dividing M. Let us consider the projective limit

$$\mathbf{Z}_S = \varprojlim_{M,\, S(M) \subset S} \mathbf{Z}/M\mathbf{Z},$$

which is taken over all positive integers M with the condition $S(M) \subset S$ with respect to the homomorphisms of reduction. From the Chinese Remainder Theorem ([Se1], [Weil]) it follows then that

$$\mathbf{Z}_S \cong \prod_{q \in S} \mathbf{Z}_q.$$

Put

$$\mathbf{Q}_S = \prod_{q \in S} \mathbf{Q}_q, \quad M_0 = \prod_{q \in S} q.$$

1.5. The Tate field. For the purposes of analysis it is convenient to embed $\mathbf{Q}_p$ into a bigger field, which is already complete both in the topological and in the algebraic sense. This field is constructed as the completion $\mathbf{C}_p = \widehat{\overline{\mathbf{Q}}}_p$ of an algebraic closure $\overline{\mathbf{Q}}_p$ of $\mathbf{Q}_p$ with respect to its single valuation satisfying the condition $|p|_p = \frac{1}{p}$. The proof of the fact $\mathbf{C}_p$ is algebraically closed is not difficult and is based on the Krasner's lemma (see [Ko1], [Ko2], [Wa]).

We will use the notations

$$\mathcal{O}_p = \{x \in \mathbf{C}_p \mid |x|_p \leq 1\}, \quad M_p = \{x \in \mathbf{C}_p \mid |x|_p < 1\}. \tag{1.13}$$

Note that the $\mathcal{O}_p$ and M_p are no longer compact, and therefore the field $\mathbf{C}_p$ is not locally compact. We also have that

$$\mathcal{O}_p/M_p = \overline{\mathbf{F}}_p \tag{1.14}$$

is the algebraic closure of $\mathbf{F}_p$.

§2. Continuous and analytic functions over a non-Archimedean field

2.1. Let K be a closed subfield of the Tate field $\mathbf{C}_p$. For a subset $W \subset K$ we consider continuous functions $f : W \to \mathbf{C}_p$. The standard examples of continuous functions are provided by polynomials, by rational functions (at points, where they are finite), and also by locally constant functions. If W is compact then for any continuous function $f : W \to \mathbf{C}_p$ and for any $\varepsilon > 0$ there exists a polynomial $h(x) \in \mathbf{C}_p[x]$ such that

$$|f(x) - h(x)|_p < \varepsilon \text{ for all } x \in W.$$

If $f(W) \subset L$ for a closed subfield L of $\mathbf{C}_p$ then $h(x)$ can be chosen so that $h(x) \in L[x]$ (see [Ko2], [Wa]).

Interesting examples of continuous p-adic functions are provided by interpolation of functions, defined on certain subsets, such as $W = \mathbf{Z}$ or $\mathbf{N}$ with $K = \mathbf{Q}_p$.

Let f be any function on non-positive integers with values in $\mathbf{Q}_p$ or in some (complete) $\mathbf{Q}_p$-Banach space. In order to extend $f(x)$ to all $x \in \mathbf{Z}_p$ we can use the interpolation polynomials

$$\binom{x}{n} = \frac{x(x-1)\cdots(x-n+1)}{n!}.$$

Then we have that $\binom{x}{n}$ is a polynomial of degree n of x, which for $x \in \mathbf{Z},\ x \geq 0$ gives the binomial coefficient. If $x \in \mathbf{Z}_p$ then x is close (in the p-adic topology) to a positive integer, hence the value of $\binom{x}{n}$ is also close to an integer, therefore $\binom{x}{n} \in \mathbf{Z}_p$.

The classical Mahler's interpolation theorem says that any continuous function $f : \mathbf{Z}_p \to \mathbf{Q}_p$ can be written in the form (see [La4], [Wa], p.52):

$$f(x) = \sum_{n=0}^{\infty} a_n \binom{x}{n}, \tag{2.1}$$

with $a_n \to 0$ (p-adically) for $n \to \infty$. For a function $f(x)$ defined for $x \in \mathbf{Z},\ x \geq 0$ one can write formally

$$f(x) = \sum_{n=0}^{\infty} a_n \binom{x}{n},$$

where the coefficients can be found from the following system of linear equations

$$f(n) = \sum_{m=0}^{n} a_m \binom{n}{m},$$

that is

$$a_m = \sum_{i=0}^{m} (-1)^{m-i} \binom{m}{i} f(i).$$

The series for $f(x)$ is always reduced to a finite sum for all $x \in \mathbf{Z},\ x \geq 0$. If $a_n \to 0$ then this series is convergent for all $x \in \mathbf{Z}_p$. As was noticed above, the inverse statement is also valid ("*Mahler's criterion*"). If convergence of a_n to zero is so fast that the series defining the coefficients of the x-expansion of $f(x)$ also converge, then $f(x)$ can be extended to an *analytic function*, see 2.2 below. Unfortunately, for an arbitrary sequence a_n with $a_n \to 0$ the attempt to use (2.1) for continuation of $f(x)$ out of the subset $\mathbf{Z}_p$ in $\mathbf{C}_p$ may fail. However, in the sequel we mostly consider analytic functions, that are defined as sums of power series.

2.2. Analytic functions and power series (see [Ko2], p.13). The well known criterion of convergence of a series $\sum_{n=0}^{\infty} a_n$ is that the partial sums $\sum_{N \leq n \leq M} a_n$ are small for large N, M with $M > N$. In view of the non-Archimedean property (1.2) in $\mathbf{C}_p$ this occurs if and only if $|a_n|_p \to 0$ or $\mathrm{ord}_p a_n \to \infty$ for $n \to \infty$. Therefore convergence of the power series $\sum_{n=0}^{\infty} a_n x^n$ depends only on $|x|_p$ but not on the precise value of x, hence there is no "conditional convergence" in this case. Thus, for any power series $\sum_{n \geq 0} a_n x^n$ we can define its radius of convergence r such that only one of the following holds:

$$\sum_{n=0}^{\infty} a_n x^n \text{ converges } \iff x \in D_0(r^-) \tag{2.2}$$

$$\sum_{n=0}^{\infty} a_n x^n \text{ converges } \iff x \in D_0(r) \tag{2.3}$$

An example of the first alternative is $\sum_{n=0}^{\infty} x^n$, where (2.2) is satisfied with $r = 1$, and an example for the second is

$$\sum_{n=0}^{\infty} p^n x^{p^n - 1};$$

here the condition (2.3) is satisfied also with $r = 1$.

The important examples of analytic functions are $\exp(x)$ and $\log(x)$, which are given as the power series

$$\exp(x) = \sum_{n=0}^{\infty} \frac{x^n}{n!}, \quad \log(1+x) = \sum_{n=1}^{\infty} \frac{(-1)^{n+1}}{n} x^n, \tag{2.4}$$

and we have that

$$\exp(x) \text{ converges on } D_0(p^{-1/(p-1)-}),\ r = p^{-1/(p-1)}, \tag{2.5}$$

$$\log(1+x) \text{ converges on } D_0(1^-),\ r = 1. \tag{2.6}$$

(see [Ko2], [B-Šaf]), so that $\exp(x)$ converges in a disc smaller than the unit disc, and $\log(1+x)$ has better convergence that $\exp(x)$.

Since the identity

$$\log(xy) = \log(x) + \log(y) \tag{2.7}$$

holds as a formal power series identity, it follows that (2.7) holds in $\mathbf{C}_p$ as long as

$$|x-1|_p < 1 \text{ and } |y-1|_p < 1.$$

In particular, since $|\zeta - 1|_p < 1$ for ζ any p^n-root of unity, we can obviously apply (2.7) to conclude that $\log\zeta = 0$. Also we have that for all $x \in D_0(p^{-1/(p-1)-})$ the following identities hold

$$\exp(\log(1+x)) = 1 + x, \quad \log(\exp(x)) = x,$$

which are deduced from the corresponding properties of the formal series and can be used for establishing isomorphisms beween certain additive and multiplicative subgroups in $\mathbf{C}_p$ and $\mathbf{C}_p^\times$: for $U = D_1(p^{\nu-}; \mathbf{Q}_p)$ (with ν as in 1.6) there are the isomorphisms

$$\exp : p^{\nu+n}\mathbf{Z}_p \xrightarrow{\sim} U^{p^n} \quad (\text{with } n \geq 0). \tag{2.8}$$

2.3. Theorem *(on analyticity of interpolation series). Let $r < p^{-1/(p-1)} < 1$ and let*

$$f(x) = \sum_{n=0}^{\infty} a_n \binom{x}{n},$$

be a series with the condition $|a_n|_p \leq Mr^n$ for some $M > 0$. Then $f(x)$ is expresible as a certain power series whose radius of convergence is not less than $R = (rp^{1/(p-1)})^{-1} > 1$ (see [Wa], p.53).

As an example let us consider the function $\langle a\rangle^x$ which is defined for $a \in \mathbf{Z}_p^\times$ by means of the decomposition $a = \omega(a)\langle a\rangle$ where $\omega(a)$ is the Teichmüller representative of a for $p > 2$ and $\omega(a) = \pm 1$ for $p = 2$ with $\omega(a) \equiv a (\mathrm{mod}\, 4)$, and the exponentiation is given by the binomial formula

$$\langle a\rangle^x = (1 + \langle a\rangle - 1)^x = \sum_{n=0}^{\infty} \binom{x}{n} (\langle a\rangle - 1)^n. \tag{2.9}$$

Since $|\langle a\rangle - 1|_p \le p^{\nu}$ we may put in the above theorem $r = p^{-\nu}$ and get that the function $\langle a\rangle^x$ is a power series in x with the radius of convergence not less than $p^{\nu-1/(p-1)} > 1$, and the following equality holds:

$$\exp(x\log\langle a\rangle) = \sum_{n=0}^{\infty}\binom{x}{n}(\langle a\rangle - 1)^n, \tag{2.10}$$

in which both parts are analytic in x and coincide for $x \in \mathbf{N}$.

2.4. Newton polygons (see [Ko2], p.21). The *Newton polygon* M_f for a power series

$$f(x) = \sum_{n=1}^{\infty} a_n x^n \in \mathbf{C}_p[[x]]$$

is defined as the convex hull of the points $(n, \mathrm{ord}_p a_n)$ (where we agree to take $\mathrm{ord}_p 0 = \infty$).

It is not hard to prove the following

Proposition. *If a segment of M_f has slope λ and horizontal length N (i.e., it extends from $(n, \mathrm{ord}_p a_n)$ to $(n+N,\ \lambda N + \mathrm{ord}_p a_n)$) then f has precisely N roots r_n with $\mathrm{ord}_p a_n = -\lambda$ (counting multiplicity).*

The following thorem is the p-adic analog of the Weierstrass Preparation Theorem (see [Ko2], p.21)

2.5. Theorem. *Let*

$$f(x) = a_m x^m + a_{m+1}x^{m+1} + \cdots \in \mathbf{C}_p[[x]],\ a_m \ne 0$$

be a power series which converges on $D_0(p^{\lambda}; \mathbf{C}_p)$. Let $(N, \mathrm{ord}_p a_N)$ be the right endpoint of the last segment of M_f with slope $\le \lambda$, if this N is infinite. Otherwise, there will be a last infinitely long segment of slope λ and only finitely many points $(n, \mathrm{ord}_p a_n)$ on that segment. In that case let N be the last such n. Then there exists a unique polynomial $h(x)$ of the form

$$b_m x^m + b_{m+1}x^{m+1} + \cdots + b_N x^N$$

with $b_m = a_m$ and a unique power series $g(x)$ which converges and does not vanish on $D_0(p^{\lambda}; \mathbf{C}_p)$ such that

$$f(x) = \frac{h(x)}{g(x)} \quad \text{on } D_0(p^{\lambda}; \mathbf{C}_p).$$

In addition M_h coincides with M_f as far as the point $(N, \mathrm{ord}_p a_N)$.

Corollary. *A power series which converges everywhere and has no zeroes is a constant.*

A simple proof of the Weierstrass Preparation Theorem for power series of the type

$$f(x) = \sum_{n=1}^{\infty} a_n x^n \in \mathcal{O}_p[[x]]$$

is based on a generalization of the Euclid algorithm (see [Man1]).

2.6. There exists another definition of the Newton polygon (see [Ko2], [V1]) of a series

$$f(x) = \sum_{n=1}^{\infty} a_n x^n \in \mathbf{C}_p[[x]]$$

Instead of the points $(n, \mathrm{ord}_p a_n)$ let us look at the lines $l_n : y = nx + \mathrm{ord}_p a_n$ with slope n and y-intercept l_n. Then $\tilde{M}_f$ is defined as the graph of the function $\min_n l_n(x)$.The x-coordinate of the points of intersection of the l_n which appear in $\tilde{M}_f$ give ord_p of the zeroes, and the difference between the slopes n of the successive l_n which appear in $\tilde{M}_f$ give the number of zeroes with given ord_p.

The first obvious application of this type of Newton polygon is that a function

$$f(x) = \sum_{n=1}^{\infty} a_n x^n \in \mathbf{C}_p[[x]]$$

is bounded in the open disc $D_0(1^-; \mathbf{C}_p)$ if and only if all the coefficients a_n are uniformly bounded.

§3. Distributions, measures, and the abstract Kummer congruences.

3.1. Distributions. Let us consider a commutative associative ring R, an R-module $\mathcal{A}$ and a profinite (i.e. compact and totally disconnected) topological space Y. Then Y is a projective limit of finite sets:

$$Y = \varprojlim_I Y_i \quad (\pi_{ij} : Y_i \to Y_j, \quad i, j \in I, \ i \geq j) \tag{3.1}$$

where I is a (partially ordered) inductive set and for $i \geq j$, $i, j \in I$ there are surjective homomorphisms $\pi_{i,j} : Y_i \to Y_j$ with the condition $\pi_{i,j} \circ \pi_{j,k} = \pi_{i,k}$ for $i \geq j \geq k$. The inductivity of I means that for any $i, j \in I$ there exist $k \in I$ with the condition $k \geq i, k \geq j$. By the universal property we have that for each $i \in I$ a unique map $\pi_i : Y \to Y_i$ is defined, which satisfies the property $\pi_{ij} \circ \pi_i = \pi_j$ (for each $i, j \in I$)

Let $\mathrm{Step}(Y, R)$ be the R-module consisting of all R-valued locally constant functions $\phi : Y \to R$.

Definition. *A distribution on Y with values in a R-module $\mathcal{A}$ is a R-linear homomorphism*

$$\mu : \mathrm{Step}(Y, R) \to \mathcal{A}. \tag{3.2}$$

For $\varphi \in \mathrm{Step}(Y, R)$ we use the notations

$$\mu(\varphi) = \int_Y \varphi d\mu = \int_Y \varphi(y) d\mu(y).$$

Each distribution μ can be defined by a system of functions $\mu^{(i)} : Y_i \to \mathcal{A}$, satisfying the following finite-additivity condition

$$\mu^{(j)}(y) = \sum_{x \in \pi_{i,j}^{-1}(y)} \mu^{(i)}(x) \quad (y \in Y_j, \ x \in Y_i). \tag{3.3}$$

In order to construct such a system it suffices to put

$$\mu^{(i)}(x) \quad = \quad \mu(\delta_{i,x}) \in \mathcal{A} \quad (x \in Y_i),$$

where $\delta_{i,x}$ is the characteristic function of the inverse image $\pi_i^{-1}(x) \subset Y$ with respect to the natural projection $Y \to Y_i$. For an arbitrary function $\varphi_j : Y_j \to R$ and $i \geq j$ we define the functions

$$\varphi_i = \varphi_j \circ \pi_{ij}, \quad \varphi = \varphi_j \circ \pi_j, \varphi \in \mathrm{Step}(Y, R), \quad \varphi_i : Y_i \xrightarrow{\pi_{ij}} Y_j \to R$$

A convenient criterion of the fact that the system of functions $\mu^{(i)} : Y_i \to \mathcal{A}$ satisfies the finite additivity condition (3.3) (and hence is associated to some distribution) is given by the following condition (*compatibility criterion*): for all $j \in I$, and $\varphi_j : Y_j \to R$ the value of the sums

$$\mu(\varphi) = \mu^{(i)}(\varphi_i) = \sum_{y_i \in Y_i} \varphi_i(y)\mu^{(i)}(y) \quad \text{is independent of } i \tag{3.4}$$
$$\text{for all large enough } i \geq j.$$

When using (3.4), it suffices to verify the condition (3.4) for some "basic" system of functions. For example, if

$$Y = G = \varprojlim_i G_i$$

is a profinite abelian group, and R is a domain containing all roots of unity of the order dividing the order of Y (which is a "supernatural number") then it is suffices to check the condition (3.4) for all characters of finite order $\chi : G \to R$, since their $R \otimes \mathbf{Q}$ -linear span coincides with the whole ring $\mathrm{Step}(Y, R \otimes \mathbf{Q})$ by the orthogonality properties for characters of a finite group [Ka3], [Maz–SD].

Example: Bernoulli distributions (see [La3]). Let M be a positive integer, $f : \mathbf{Z} \to \mathbf{C}$ is a periodic function with the period M (i.e. $f(x + M) = f(x)$, $f : \mathbf{Z}/M\mathbf{Z} \to \mathbf{C}$). The generalized Bernoulli number (see [Le1], [Šaf1]) $B_{k,f}$ is defined as $k!$ times the coefficient by t^k in the expansion in t of the rational quotient

$$\sum_{a=0}^{M-1} \frac{f(a)te^{at}}{e^{Mt} - 1},$$

that is,

$$\sum_{k=0}^{\infty} \frac{B_{k,f}}{k!} t^k = \sum_{a=0}^{M-1} \frac{f(a)te^{at}}{e^{Mt} - 1}. \tag{3.5}$$

Now let us consider the profinite ring

$$Y = \mathbf{Z}_S = \varprojlim_M \mathbf{Z}/M\mathbf{Z} \quad (S(M) \subset S),$$

the projective limit being taken over the set of all positive integers M with support $S(M)$ in a fixed finite set S of prime numbers. Then the periodic function $f : \mathbf{Z}/M\mathbf{Z} \to \mathbf{C}$

with $S(M) \subset S$ can be regarded as an element of $\mathrm{Step}(Y, \mathbf{C})$. We claim that there exists a distribution $E_k : \mathrm{Step}(Y, \mathbf{C}) \to \mathbf{C}$ which is uniquely determined by the condition

$$E_k(f) = B_{k,f} \quad \text{for all} \quad f \in \mathrm{Step}(Y, \mathbf{C}). \tag{3.6}$$

In order to prove the existence of this distribution we use the above criterion (3.4) and check that for every $f \in \mathrm{Step}(Y, \mathbf{C})$ the right hand side in (3.6) (i.e. $B_{k,f}$) does not depend on the choice of a period M of the function f. It follows directly from the definition (3.5); however we give here a different proof which is based on an interpretation of the numbers $B_{k,f}$ as certain special values of L-functions.

For the function $f : \mathbf{Z}/M\mathbf{Z} \to \mathbf{C}$ let

$$L(s, f) = \sum_{n=1}^{\infty} f(n) n^{-s}$$

be the corresponding L-series which is absolutely convergent for all s with $\mathrm{Re}(s) > 1$ and admits an analytic continuation over all $s \in \mathbf{C}$. For this series we have that

$$L(1-k, f) = -\frac{B_{k,f}}{k}. \tag{3.7}$$

For example, if $f \equiv 1$ is the constant function with the period $M = 1$ then we have that

$$\zeta(1-k) = -\frac{B_k}{k}, \quad \sum_{k=0}^{\infty} \frac{B_k}{k!} t^k = \frac{t}{e^t - 1},$$

B_k being the Bernoulli number. The formula (3.7) is established by means of the contour integral discovered by Riemann (see [La3], ch.XXI). This formula apparently implies the desired independence of $B_{k,f}$ on the choice of M. We note also that if $K \subset \mathbf{C}$ is an arbitrary subfield, and $f(Y) \subset K$ then we have from the formula (3.5) that $B_{k,f} \in K$ hence the distribution E_k is a K-valued distribution on Y.

3.2. Measures. Let R be a topological ring, and $\mathcal{C}(Y, R)$ be the topological module of all R-valued functions on a profinite set Y.

Definition. *A measure on Y with values in the topological R-module $\mathcal{A}$ is a continuous homomorphism of R-modules*

$$\mu : \mathcal{C}(Y, R) \to \mathcal{A}.$$

The restriction of μ to the R-submodule $\mathrm{Step}(Y, R) \subset \mathcal{C}(Y, R)$ defines a distribution which we denote by the same letter μ, and the measure μ is uniquely determined by the corresponding distribution since the R-submodule $\mathrm{Step}(Y, R)$ is dense in $\mathcal{C}(Y, R)$. The last statement expresses the well known fact about the uniform continuity of a continuous function over a compact topological space.

Now we consider any closed subring R of the Tate field $\mathbf{C}_p$, $R \subset \mathbf{C}_p$, and let $\mathcal{A}$ be a complete R-module with topology given by a norm $|\cdot|_{\mathcal{A}}$ on $\mathcal{A}$ compatible with the norm $|\cdot|_p$ on $\mathbf{C}_p$ so that the following conditions are satisfied:

for $x \in \mathcal{A}$ the equality $|x|_{\mathcal{A}} = 0$ is equivalent to $x = 0$,

for $a \in R$, $x \in \mathcal{A}$ $\quad |ax|_{\mathcal{A}} = |a|_p |x|_{\mathcal{A}}$,

for all $x, y \in \mathcal{A}$ $\quad |x + y|_{\mathcal{A}} \le \max(|x|_{\mathcal{A}}, |y|_{\mathcal{A}})$.

Then the fact that a distribution (a system of functions $\mu^{(i)} : Y_i \to \mathcal{A}$) gives rise to a $\mathcal{A}$-valued measure on Y is equivalent to the condition that the system $\mu^{(i)}$ is bounded, i.e. for some constant $B > 0$ and for all $i \in I$, $x \in Y_i$ the following uniform estimate holds:

$$|\mu^{(i)}(x)|_{\mathcal{A}} < B. \tag{3.8}$$

This criterion is an easy consequence of the non-Archimedean property

$$|x + y|_{\mathcal{A}} \le \max(|x|_{\mathcal{A}}, |y|_{\mathcal{A}})$$

of the norm $|\cdot|_{\mathcal{A}}$ (see [Man4], [V1]).

In particular if $\mathcal{A} = R = \mathcal{O}_p = \{x \in \mathbf{C}_p |\ |x|_p \le 1\}$ is the subring of integers in the Tate field $\mathbf{C}_p$ then the set of $\mathcal{O}_p$-valued distributions on Y coincides with $\mathcal{O}_p$-valued measures (in fact, both sets are R-algebras with multiplication defined by convolution, see §4).

Below we give some examples of measures based on the following important criterion of existence of a measure with given properties.

3.3. Proposition (The abstract Kummer congruences) (see [Ka3], p.258). *Let $\{f_i\}$ be a system of continuous functions $f_i \in \mathcal{C}(Y, \mathcal{O}_p)$ in the ring $\mathcal{C}(Y, \mathcal{O}_p)$ of all continuous functions on the compact totally disconnected group Y with values in the ring of integers $\mathcal{O}_p$ of $\mathbf{C}_p$ such that $\mathbf{C}_p$-linear span of $\{f_i\}$ is dense in $\mathcal{C}(Y, \mathbf{C}_p)$. Let also $\{a_i\}$ be any system of elements $a_i \in \mathcal{O}_p$. Then the existence of an $\mathcal{O}_p$ -valued measure μ on Y with the property*

$$\int_Y f_i \, d\mu = a_i$$

is equivalent to the following congruences: for an arbitrary choice of elements $b_i \in \mathbf{C}_p$ almost all of which vanish

$$\sum_i b_i f_i(y) \in p^n \mathcal{O}_p \text{ for all } y \in Y \text{ implies } \sum_i b_i a_i \in p^n \mathcal{O}_p. \tag{3.9}$$

Remark. *Since $\mathbf{C}_p$ -measures are characterized as bounded $\mathbf{C}_p$-valued distributions, every $\mathbf{C}_p$-measure on Y becomes a $\mathcal{O}_p$-valued measure after multiplication by some non-zero constant.*

The proof of proposition 3.3. The nessecity is obvious since

$$\sum_i b_i a_i = \int_Y (p^n \mathcal{O}_p\text{−valued function })d\mu =$$

$$p^n \int_Y (\mathcal{O}_p\text{−valued function })d\mu \in p^n \mathcal{O}_p.$$

In order to prove the sufficiency we need to construct a measure μ from the numbers a_i. For a function $f \in \mathcal{C}(Y, \mathcal{O}_p)$ and a positive integer n there exist elements $b_i \in \mathbf{C}_p$ such that only a finite number of b_i does not vanish, and

$$f - \sum_i b_i f_i \in p^n \mathcal{C}(Y, \mathcal{O}_p)$$

according to the density of the $\mathbf{C}_p$-span of $\{f_i\}$ in $\mathcal{C}(Y, \mathbf{C}_p)$. By the assumption (3.9) the value $\sum_i a_i b_i$ belongs to $\mathcal{O}_p$ and is well defined modulo p^n (i.e. does not depend on the choice of b_i). Following N.M.Katz [Ka3], we denote this value by "$\int_Y f\,d\mu \mod p^n$". Then we have that the limit procedure

$$\int_Y f\,d\mu = \lim_{n\to\infty} \text{“}\int_Y f\,d\mu \mod p^n\text{”} \in \varprojlim_n \mathcal{O}_p/p^n\mathcal{O}_p = \mathcal{O}_p,$$

gives the measure μ.

3.4. The S-adic Mazur measure. Let $c > 1$ be a positive integer coprime to

$$M_o = \prod_{q\in S} q$$

with S being a fixed set of prime numbers. Using the criterion 3.3 we show that the $\mathbf{Q}$-valued distribution defined by the formula

$$E_k^c(f) = E_k(f) - c^k E_k(f_c), \quad f_c(x) = f(cx), \tag{3.10}$$

turns out to be a measure where $E_k(f)$ are defined in 3.1, $f \in \mathrm{Step}(Y, \mathbf{Q})$ and the field $\mathbf{Q}$ is regarded as subfield of $\mathbf{C}_p$. Define the generalized Bernoulli polynomials $B_{k,f}^{(M)}(X)$:

$$\sum_{k=0}^{\infty} B_{k,f}^{(M)}(X)\frac{t^k}{k!} = \sum_{a=0}^{M-1} f(a)\frac{t\,e^{(a+X)t}}{e^{Mt}-1} \tag{3.11}$$

and the generalized sums of powers

$$S_{k,f}(M) = \sum_{a=0}^{M-1} f(a)a^k.$$

Then the definition (3.11) formally implies that

$$\frac{1}{k}[B_{k,f}^{(M)}(M) - B_{k,f}^{(M)}(0)] = S_{k-1,f}(M), \tag{3.12}$$

and also we see that

$$B_{k,f}^{(M)}(X) = \sum_{i=0}^{k} \binom{k}{i} B_{i,f}X^{k-i} = B_{k,f} + kB_{k-1,f}X + \cdots + B_{0,f}X^k \tag{3.13}$$

The last identity can be rewritten symbolically as

$$B_{k,f}(X) = (B_f + X)^k.$$

The equality (3.12) enabels us to calculate the (generalized) sums of powers in terms of the (generalized) Bernoulli numbers. In particular this equality implies that the

Bernoulli numbers $B_{k,f}$ can be obtained by the following p-adic limit procedure ([La3], Chapter XIII):

$$B_{k,f} = \lim_{n\to\infty} \frac{1}{Mp^n} S_{k,f}(Mp^n) \quad (p-\text{adic limit}), \tag{3.14}$$

where f is a $\mathbf{C}_p$-valued function on $Y = \mathbf{Z}_S$. Indeed, if we replace M in (3.12) by Mp^n with growing n and let D be the common denominator of all coefficients of the polynomial $B_{k,f}^{(M)}(X)$. Then we have from (3.13) that

$$\frac{1}{k}[B_{k,f}^{(Mp^n)}(M) - B_{k,f}^{(M)}(0)] \equiv B_{k-1,f}(Mp^n) \bmod \frac{1}{kD}p^2 n. \tag{3.15}$$

The proof of (3.14) is accomplished by division of (3.15) by Mp^n and by application of the formula (3.12).

Now we can directly show that the distributions E_k^c defined by (3.10) are in fact bounded measures. If we use (3.9) and take the functions $\{f_i\}$ to be all of the functions in $\mathrm{Step}(Y, \mathcal{O}_p)$. Let $\{b_i\}$ be a system of elements $b_i \in \mathbf{C}_p$ such that for all $y \in Y$ the congruence

$$\sum_i b_i f_i(y) \equiv 0 (\bmod p^n) \tag{3.16}$$

holds. Set $f = \sum_i b_i f_i$ and assume (without loss of generality) that the number n is large enough so that for all i with with $b_i \neq 0$ the congruence

$$B_{k,f_i} \equiv \frac{1}{Mp^n} S_{k,f_i}(Mp^n) \bmod p^n \tag{3.17}$$

is valid in accordance with (3.14). Then we see that

$$B_{k,f} \equiv (Mp^n)^{-1} \sum_i \sum_{a=0}^{Mp^n-1} b_i f_i(a) a^k \bmod p^n, \tag{3.18}$$

hence we get by definition (3.10):

$$\begin{aligned} &E_k^c(f) = B_{k,f} - c^k B_{k,f_c} \equiv \\ &(Mp^n)^{-1} \sum_i \sum_{a=0}^{Mp^n-1} b_i [f_i(a) a^k - f_i(ac)(ac)^k] \bmod p^n. \end{aligned} \tag{3.19}$$

Let

$$a_c \in \{0, 1, ..., Mp^n - 1\}, \quad a_c \equiv ac \bmod Mp^n,$$

then the map $a \mapsto a_c$ is well defined and acts as a permutation of the set

$$\{0, 1, ..., Mp^n - 1\},$$

hence (3.19) is equivalent to the congruence

$$E_k^c(f) = B_{k,f} - c^k B_{k,f_c} \equiv \sum_i \frac{a_c^k - (ac)^k}{Mp^n} \sum_{a=0}^{Mp^n-1} b_i f_i(a) a^k \bmod p^n. \tag{3.20}$$

Now the assumption (3.16) formally implies that $E_k^c(f) \equiv 0 \bmod p^n$, completing the proof of the abstract Kummer congruences and the construction of measures E_k^c.

Remark. *The formula (3.20) also implies that for all $f \in \mathcal{C}(Y, \mathbf{C}_p)$ the following hold*

$$E_k^c(f) = kE_1^c(x_p^{k-1} f) \tag{3.21}$$

where $x_p : Y \to \mathbf{C}_p \in \mathcal{C}(Y, \mathbf{C}_p)$ is the composition of the projection $Y \to \mathbf{Z}_p$ and the embedding $\mathbf{Z}_p \hookrightarrow \mathbf{C}_p$.

Indeed if we put $a_c = ac + Mp^n t$ for some $t \in \mathbf{Z}$ then we see that

$$a_c^k - (ac)^k = (ac + Mp^n t)^k - (ac)^k \equiv kMp^n t (ac)^{k-1} \mod (Mp^n)^2),$$

and we get that in (3.20)

$$\frac{a_c^k - (ac)^k}{Mp^n} \equiv k(ac)^{k-1} \frac{a_c - ac}{Mp^n} (\bmod Mp^n).$$

The last congruence is equivalent to saying that the abstract Kummer congruences (3.9) are satisfied by all functions of the type $x_p^{k-1} f_i$ and by the measure E_1^c with $f_i \in \mathrm{Step}(Y, \mathbf{C}_p)$ establishing the identity (3.21).

§4. Iwasawa algebra and the non-Archimedean Mellin transform.

4.1. The domain of definition of the non-Archimedean zeta functions. In the classical case the set on which zeta functions are defined is the set of complex numbers $\mathbf{C}$ which can be regarded equally as the set of all continuous characters (more precisely, quasicharacters) via the following isomorphism:

$$\mathbf{C} \xrightarrow{\sim} \mathrm{Hom}_{\mathrm{contin}}(\mathbf{R}_+^\times, \mathbf{C}^\times); \quad s \mapsto (y \mapsto y^s). \tag{4.1}$$

The construction which associates to a function $h(y)$ on $\mathbf{R}_+^\times$ (with certain growth conditions for $y \to \infty$ and $y \to 0$) the integral

$$L_h(s) = \int_{\mathbf{R}_+^\times} h(y)\, y^s \frac{dy}{y}$$

(which probably converges not for all values of s) is called the *Mellin transform.* For example, if $\zeta(s) = \sum_{n=1}^\infty n^{-s}$ is the Riemann zeta function, then the function $\zeta(s)\Gamma(s)$ is the Mellin transform of the function $h(y) = 1/(1 - e^{-y})$:

$$\zeta(s)\Gamma(s) = \int_0^\infty \frac{1}{1 - e^{-y}} y^s \frac{dy}{y}, \tag{4.2}$$

so that the integral and the series are absolutely convergent for $\mathrm{Re}(s) > 1$. For an arbitrary function of the type

$$f(z) = \sum_{n=1}^{\infty} a(n) e^{2\pi i n z}$$

with $z = x + iy \in \mathfrak{H}$ in the upper half plane $\mathfrak{H}$ and with the growth condition $a(n) = \mathcal{O}(n^c)$ $(c > 0)$ on its Fourier coefficients, we see that the zeta function

$$L(s, f) = \sum_{n=1}^{\infty} a(n) n^{-s},$$

essentially coincides with the Mellin transform of $f(z)$, that is

$$\frac{\Gamma(s)}{(2\pi)^s} L(s, f) = \int_0^{\infty} f(iy) y^s \frac{dy}{y}. \tag{4.3}$$

Both sides of the equality (4.3) converge absolutely for $\mathrm{Re}(s) > 1 + c$. The identities (4.2) and (4.3) are immediately deduced from the well known integral representation for the gamma-function

$$\Gamma(s) = \int_0^{\infty} e^{-y} y^s \frac{dy}{y}, \tag{4.4}$$

where $\frac{dy}{y}$ is a measure on the group $\mathbf{R}_+^{\times}$ which is invariant under the group translations (Haar measure). The integral (4.4) is absolutely convergent for $\mathrm{Re}(s) > 0$ and it can be interpreted as the integral of the product of an additive character $y \to e^{-y}$ of the group $\mathbf{R}^{(+)}$ restricted to $\mathbf{R}_+^{\times}$, and the multiplicative character $y \mapsto y^s$ with respect to the Haar measure on the group $\mathbf{R}_+^{\times}$.

In the theory of non-Archimedean integration one considers the group $\mathbf{Z}_S^{\times}$ (the group of units of the S-adic completion of the ring of integers $\mathbf{Z}$) instead of the group $\mathbf{R}_+^{\times}$, and the Tate field $\mathbf{C}_p = \widehat{\overline{\mathbf{Q}}}_p$ (the completion of an algebraic closure of $\mathbf{Q}_p$) instead of the complex field $\mathbf{C}$. The domain of definition of p-adic zeta functions is the p-adic analytic Lie group

$$X_S = \mathrm{Hom}_{\mathrm{contin}}(\mathbf{Z}_S^{\times}, \mathbf{C}_p^{\times}) = X(\mathbf{Z}_S^{\times}), \tag{4.5}$$

where

$$\mathbf{Z}_S^{\times} \cong \oplus_{q \in S} \mathbf{Z}_q^{\times},$$

and the symbol

$$X(G) = \mathrm{Hom}_{\mathrm{contin}}(G, \mathbf{C}_p^{\times}) \tag{4.6}$$

denotes the functor of all p-adic characters of a topological group G (see[V1]).

4.2. The analytic structure of X_S. Let us consider in detail the structure of the topological group X_S. Define

$$U_p = \{x \in \mathbf{Z}_p^{\times} | x \equiv 1 \,(\mathrm{mod}\, p^{\nu})\}$$

where $\nu = 1$ or $\nu = 2$ according as $p > 2$ or $p = 2$. Then we have the natural decomposition

$$X_S = X((\mathbf{Z}/p^\nu\mathbf{Z})^\times \times \prod_{q\neq p} \mathbf{Z}_q^\times) \times X(U_p). \tag{4.7}$$

The analytic structure on $X(U_p)$ is defined by the following isomorphism (which is equivalent to a special choice of a local parameter):

$$\varphi : X(U_p) \xrightarrow{\sim} T = \{z \in \mathbf{C}_p^\times \mid |t-1|_p < 1\},$$

where $\varphi(x) = x(1+p^\nu)$, $1+p^\nu$ being a topological generator of the multiplicative group $U_p \cong \mathbf{Z}_p$. An arbitrary character $\chi \in X_S$ can be uniquely represented in the form $\chi = \chi_0 \cdot \chi_1$ where χ_0 is trivial on the component U_p, and χ_1 is trivial on the other component

$$(\mathbf{Z}/p^\nu\mathbf{Z})^\times \times \prod_{q\neq p} \mathbf{Z}_q^\times.$$

The character χ_0 is called the *tame component*, and χ_1 the *wild component* of the character χ. We denote by the symbol $\chi_{(t)}$ the (wild) character which is uniquely determined by the condition

$$\chi_{(t)}(1+p^\nu) = t$$

with $t \in \mathbf{C}_p$, $|t|_p < 1$.

In some cases it is convenient to use another local coordinate s which is analogous to the classical argument s of the Dirichlet series:

$$\mathcal{O}_p \ni s \mapsto \chi^{(s)}((1+p^\nu)^\alpha) = (1+p^\nu)^{\alpha s} = \exp(\alpha s \log(1+p^\nu)).$$

The character $\chi^{(s)}$ is defined only for such s for which the series exp is p-adically convergent. In this domain of values of the argument we have that $t = (1+p^\nu)^s - 1$. But, for example, for $(1+t)^{p^n} = 1$ there is certainly no such value of s, so that the s-coordinate parametrizes a smaller neighborhood of the trivial character then the t-coordinate (which parametrizes all wild characters) (see [Man4], [Man6]).

Recall that an analytic function $F : T \to \mathbf{C}_p$ is defined as the sum of a series of the type $\sum_{i=0}^{\infty} a_i(t-1)^i$ ($a_i \in \mathbf{C}_p$), which is assumed to be absolutely convergent for all $t \in T$. The notion of an analytic function is then obviously extended to the whole group X_S by shifts. The function

$$F(t) = \sum_{i=0}^{\infty} a_i(t-1)^i$$

is bounded on T iff all its coefficients a_i are universally bounded. This last fact can be easily deduced for example from the basic properties of the Newton polygon of the series $F(t)$ (see [Ko2], [V1], [V2]). If we apply to these series the Weierstrass preparation theorem ([Ko1], [La3], [Man1] and 2.5) we see that in this case the function F has only a finite number of zeroes on T (if it is not identically zero). In particular, consider the torsion subgroup $X_S^{\mathrm{tors}} \subset X_S$. This subgroup is discrete in X_S and its elements $v \in X_S^{\mathrm{tors}}$ can be obviously identified with primitive Dirichlet characters χ mod M

such that the support $S(\chi) = S(M)$ of the conductor of χ is contained in S. This identification is provided by a fixed embedding

$$i_p : \overline{\mathbf{Q}}^\times \hookrightarrow \mathbf{C}_p^\times$$

if we note that each character $\chi \in X_S^{\mathrm{tors}}$ can be factored through some finite factor group $(\mathbf{Z}/M\mathbf{Z})^\times$:

$$\chi : \mathbf{Z}_S^\times \to (\mathbf{Z}/M\mathbf{Z})^\times \to \overline{\mathbf{Q}}^\times \overset{i_p}{\hookrightarrow} \mathbf{C}_p^\times,$$

and the smallest number M with the above condition is the conductor of $\chi \in X_S^{\mathrm{tors}}$.

The symbol x_p will denote the composition of the natural projection $\mathbf{Z}_S^\times \to \mathbf{Z}_p^\times$ and of the natural embedding $\mathbf{Z}_p^\times \to \mathbf{C}_p^\times$, so that $x_p \in X_S$ and all integers k can be considered as the characters $x_p^k : y \mapsto y^k$.

Let us consider a bounded $\mathbf{C}_p$-analytic function F on X_S. The above statement about zeroes of bounded $\mathbf{C}_p$-analytic functions implies now that the function F is uniquely determined by its values $F(\chi_0\chi)$, where χ_0 is a fixed character and χ runs through all elements $\chi \in X_S^{\mathrm{tors}}$ with possible exclusion of a finite number of characters in each analyticity component of the decomposition (4.7). This condition is satisfied, for example, by the set of characters $\chi \in X_S^{\mathrm{tors}}$ with the S-complete conductor (i.e. such that $S(\chi) = S$), and even for the smaller set of characters which is obtained by imposing the additional assumption that the character χ^2 is not trivial (see [Man4], [Man6], [V1]).

4.3. The non-Archimedean Mellin transform. Let μ be a (bounded) $\mathbf{C}_p$-valued measure on $\mathbf{Z}_S^\times$. Then the *non-Archimedean Mellin transform* of a measure μ is defined by

$$L_\mu(x) = \mu(x) = \int_{\mathbf{Z}_S^\times} x \, d\mu \quad (x \in X_S), \tag{4.8}$$

which represents a bounded $\mathbf{C}_p$-analytic function

$$L_\mu : X_S \to \mathbf{C}_p. \tag{4.9}$$

Indeed, the boundeness of the function L_μ is obvious since all characters $x \in X_S$ take values in $\mathcal{O}_p$ and μ also is bounded. The analyticity of this function expresses a general property of the integral (4.8), namely, that it depends analytically on the parameter $x \in X_S$. However, we give below a pure algebraic proof of this fact which is based on a description of the Iwasawa algebra. This description will also imply that every bounded $\mathbf{C}_p$-analytic function on X_S is the Mellin transform of a certain measure μ.

4.4. The Iwasawa algebra (see [La3], chapter XII). Let $\mathcal{O}$ be a closed subring in $\mathcal{O}_p = \{z \in \mathbf{C}_p \mid |z|_p \le 1\}$,

$$G = \varprojlim_i G_i \quad (i \in I)$$

be a profinite group. Then the canonical homomorphism $G_i \overset{\pi_{ij}}{\longleftarrow} G_j$ induces a homomorphism of the corresponding group rings

$$\mathcal{O}[G_i] \longleftarrow \mathcal{O}[G_j].$$

Then *the completed group ring* $\mathcal{O}[[G]]$ is defined as the projective limit

$$\mathcal{O}[[G]] = \varprojlim_i \mathcal{O}[[G_i]] \quad (i \in I)$$

Let us consider also the set $\mathcal{D}istr\,(G, \mathcal{O})$ of all $\mathcal{O}$-valued distributions on G which itself is an $\mathcal{O}$-module and a ring with respect to multiplication given by *the convolution of distributions*, which is defined in terms of families of functions

$$\mu_1^{(i)}, \mu_2^{(i)} : G_i \to \mathcal{O}$$

(see the previous section) as follows:

$$(\mu_1 * \mu_2)^{(i)}(y) = \sum_{y=y_1 y_2} \mu_1^{(i)}(y_1)\mu_2^{(i)}(y_2) \quad (y_1,\ y_2 \in G_i) \tag{4.10}$$

Recall also that the $\mathcal{O}$-valued distributions are identified with $\mathcal{O}$-valued measures. Now we describe an isomorphism of $\mathcal{O}$-algebras $\mathcal{O}[[G]]$ and $\mathcal{D}istr\,(G, \mathcal{O})$. In the case when $G = \mathbf{Z}_p$ the algebra $\mathcal{O}[[G]]$ is called *the Iwasawa algebra.*

4.5.Theorem. *(a) Under the notation of 4.4 there is the canonical isomorphism of $\mathcal{O}$-algebras*

$$\mathcal{D}istr\,(G, \mathcal{O}) \xrightarrow{\sim} \mathcal{O}[[G]]; \tag{4.11}$$

(b) If $G = \mathbf{Z}_p$ then there is an isomorphism

$$\mathcal{O}[[G]] \xrightarrow{\sim} \mathcal{O}[[X]], \tag{4.12}$$

where $\mathcal{O}[[X]]$ is the ring of formal power series in X over $\mathcal{O}$. The isomorphism (4.12) depends on a choice of the topological generator of the group $G = \mathbf{Z}_p$.

4.6. Formulas for coefficients of power series. We noticed above that the theorem 4.5 would imply a description of $\mathbf{C}_p$-analytic bounded functions on X_S in terms of measures. Indeed, these functions are defined on analyticity components of the decomposition (4.7) as certain power series with p-adically bounded coefficients, that is, power series, whose coefficients belong to $\mathcal{O}_p$ after multiplication by some non-zero constant from $\mathbf{C}_p^\times$. Formulas for coefficients of these series can be also deduced from the proof of the theorem. However, we give a more direct computation of these coefficients in terms of the corresponding measures. Let us consider the component aU_p of the set $\mathbf{Z}_S^\times$ where

$$a \in (\mathbf{Z}/p^\nu \mathbf{Z})^\times \times \prod_{q \neq p} \mathbf{Z}_q^\times,$$

and let $\mu_a(x) = \mu(ax)$ be the corresponding measure on U_p defined by restriction of μ to the subset $aU_p \subset \mathbf{Z}_S^\times$. Consider the isomorphism $U_p \cong \mathbf{Z}_p$ given by

$$y = \gamma^x \ (x \in \mathbf{Z}_p,\ y \in U_p)$$

with some choice of the generator γ of U_p (for example, we can take $\gamma = 1 + p^\nu$). Let μ'_a be the corresponding measure on $\mathbf{Z}_p$. Then this measure is uniquely determined by values of the integrals

$$\int_{\mathbf{Z}_p} \binom{x}{i} d\mu'_a(x) = a_i, \tag{4.13}$$

with the interpolation polynomials $\binom{x}{i}$, since the $\mathbf{C}_p$-span of the family

$$\left\{\binom{x}{i}\right\} \quad (i \in \mathbf{Z}, i \geq 0)$$

is dense in $\mathcal{C}(\mathbf{Z}_p, \mathcal{O}_p)$ according to the Mahler's interpolation theorem for continuous functions on $\mathbf{Z}_p$ (see 2.1 and [Mah]). Then from the basic properties of the interpolation polynomials it follows that

$$\sum_i b_i \binom{x}{i} \equiv 0(\mathrm{mod} p^n) \text{ (for all } x \in \mathbf{Z}_p) \Longrightarrow b_i \equiv 0(\mathrm{mod} p^n).$$

We can now apply the abstract Kummer congruences (see 3.3), which imply that for arbitrary choice of numbers $a_i \in \mathcal{O}_p$ there exists a measure with the property (4.13).

On the other hand we state that the Mellin transform L_{μ_a} of the measure μ_a is given by the power series $F_a(t)$ with coefficients as in (4.13), that is

$$\int_{U_p} \chi_{(t)}(y)\, d\mu(ay) = \sum_{i=0}^{\infty} \left(\int_{\mathbf{Z}_p} \binom{x}{i} d\mu'_a(x)\right)(t-1)^i \tag{4.14}$$

for all wild characters of the form $\chi_{(t)}$, $\chi_{(t)}(\gamma) = t$, $|t-1|_p < 1$. It suffices to show that (4.14) is valid for all characters of the type $y \mapsto y^m$, where m is a positive integer. In order to do this we use the binomial expansion

$$\gamma^{mx} = (1 + (\gamma^m - 1))^x = \sum_{i=0}^{\infty} \binom{x}{i} (\gamma^m - 1)^i,$$

which implies that

$$\int_{U_p} y^m\, d\mu(ay) = \int_{\mathbf{Z}_p} \gamma^{mx} d\mu'_a(x) = \sum_{i=0}^{\infty} \left(\int_{\mathbf{Z}_p} \binom{x}{i} d\mu'_a(x)\right)(\gamma^m - 1)^i,$$

establishing (4.14).

4.7. Example. The S-adic Mazur measure and the non-Archimedean Kubota – Leopoldt zeta function (see [La3], [Ku-Le], [Le2], [Wa]). Let us first consider a positive integer $c \in \mathbf{Z}_S^\times \cap \mathbf{Z}$, $c > 1$ coprime to all primes in S. Then for each complex number $s \in \mathbf{C}$ there exists a complex distribution μ_s^c on $G_S = \mathbf{Z}_S^\times$ which is uniquely determined by the following condition

$$\mu_s^c(\chi) = (1 - \chi^{-1}(c) c^{-1-s}) L_{M_0}(-s, \chi), \tag{4.15}$$

where $M_0 = \prod_{q \in S} q$ (see 3.1). Moreover, the right hand side of (4.15) is holomorphic for all $s \in \mathbf{C}$ including $s = -1$. If s is an integer and $s \geq 0$ then according to criterion

(3.4) the right-hand side of (4.15) belongs to the field

$$\mathbf{Q}(\chi) \subset \mathbf{Q}^{\mathrm{ab}} \subset \overline{\mathbf{Q}}$$

generated by values of the character χ, and we get a distribution with values in $\mathbf{Q}^{\mathrm{ab}}$. If we now apply to (4.15) the fixed embedding $i_p : \overline{\mathbf{Q}} \hookrightarrow \mathbf{C}_p$ we get a $\mathbf{C}_p$-valued distribution $\mu^{(c)} = i_p(\mu_0^c)$ which turns out to be an $\mathcal{O}_p$-measure in view of 3.3, and the following equality holds

$$\mu^{(c)}(\chi x_p^r) = i_p(\mu_r^c(\chi)).$$

This identity relates the special values of the Dirichlet L-functions at different non-positive points. The function

$$L(x) = (1 - c^{-1}x(c)^{-1})^{-1} L_{\mu^{(c)}}(x) \quad (x \in X_S) \tag{4.16}$$

is well defined and it is holomorphic on X_S with the exception of a simple pole at the point $x = x_p \in X_S$. This function is called the *non-Archimedean zeta-function of Kubota – Leopoldt*. The corresponding measure $\mu^{(c)}$ will be called the *S-adic Mazur measure*.

§5. Complex valued distributions, associated with Euler products

5.1. In this section we give a general construction of distributions, attached to rather arbitrary Euler products. This construction provides a generalization of measures, which were first introduced by Yu.I.Manin [Man4], B.Mazur and H.P.F.Swinnerton – Dyer [Maz-SD]. Our construction [Pa5], [Pa9] was already successfully used in several problems concerning the p-adic analytic interpolation of Dirichlet series [Ar], [Co-Schm], [Sch].

5.2. Let S be a fixed finite set of primes and

$$\mathcal{D}(s) = \sum_{n=1}^{\infty} a_n n^{-s} \quad (s,\ a_n \in \mathbf{C}) \tag{5.1}$$

be a Dirichlet series with the following multiplicativity property of its coefficients a_n:

$$\mathcal{D}(s) = \prod_{q \in S} F_q(q^{-s})^{-1} \sum_{\substack{n=1 \\ (n,S)=1}}^{\infty} a_n n^{-s}, \tag{5.2}$$

where the condition $(n, S) = 1$ means that n is not divisible by any prime in S, and $F_q(X)$ are polynomials with the constant term equal to 1:

$$F_q(X) = 1 + \sum_{i=1}^{m_q} A_{q,i} X^i. \tag{5.3}$$

We assume also that the series (5.1) is absolutely convergent in some right half plane $\mathrm{Re}(s) > 1 + c \quad (c \in \mathbf{R})$. This assumption is satisfied in most cases, for example, when the coefficients a_n satisfy the estimate $|a_n| = \mathcal{O}(n^c)$. For a Dirichlet character $\chi : (\mathbf{Z}/M\mathbf{Z})^\times \to \mathbf{C}^\times$ modulo $M \geq 1$ the twisted Dirichlet series is defined by

$$\mathcal{D}(s,\chi) = \sum_{n=1}^{\infty} \chi(n) a_n n^{-s}. \tag{5.4}$$

For all $s \in \mathbf{C}$ such that the series (5.1) is absolutely convergent let us define the function $P_s : \mathbf{Q} \to \mathbf{C}$ by the equality

$$P_s(x) = \sum_{n=1}^{\infty} e(nx) a_n n^{-s} \quad (e(x) = \exp(2\pi i x)). \tag{5.5}$$

Using the functions (5.5) we construct distributions μ_s on the compact group $\mathbf{Z}_S^\times$ such that for every primitive Dirichlet charactder χ viewed as a homomorphism $\chi : \mathbf{Z}_S^\times \to \mathbf{C}^\times$ the value of the series $\mathcal{D}(s,\overline{\chi})$ at s with $\mathrm{Re}(s) > 1+c$ is canonically expressed in terms of the integral

$$\int_{\mathbf{Z}_S^\times} \chi \, d\mu_s \stackrel{\text{def}}{=} \sum_{\substack{a \bmod M \\ (a,S)=1}} \chi(a)\, \mu_s(a + (M)).$$

Let for every $q \in S$, $\alpha(q)$ denote a fixed root of the inverse polynomial

$$X^{m_q} F_q(X^{-1}) = X^{m_q} + \sum_{i=1}^{m_q} A_{q,i} X^{m_q - i}$$

(that is, an inverse root of $F_q(X)$). We suppose that $\alpha(q) \neq 0$ for every $q \in S$, and let us extend the definition of numbers $\alpha(n)$ to all positive integers whose support is contained in S (by multiplicativity):

$$\alpha(n) = \prod_{q \in S} \alpha(q)^{\mathrm{ord}_q n} \quad (S(q) \subset S).$$

Let us define an auxiliary polynomial

$$H_q(X) = 1 + \sum_{i=1}^{m_q - 1} B_{q,i} X^i \tag{5.6}$$

by means of the relation

$$F_q(X) = (1 - \alpha(q)X) H_q(X), \tag{5.7}$$

which implies the identities

$$B_{q,i} = -\sum_{j=i+1}^{m_q} A_{q,j} \alpha(q)^{i-j} \quad (i = 1, \cdots, m_q - 1) \tag{5.8}$$

for the coefficients of the polynomial (5.6). Let us also introduce the following finite Euler product

$$\sum_{S(n)\subset S} B^S(n)n^{-s} = \prod_{q\in S} H_q(q^{-s}), \tag{5.9}$$

in which the coefficients $B^S(n)$ are given by means of (5.8), namely,

$$B^S(n) = \prod_{q\in S} B^S(q^{\mathrm{ord}_q n}) \quad (S(n) \subset S),$$

with

$$B^S(q^i) = \begin{cases} B_{q,i}, & \text{for } i < m_q \\ 0, & \text{otherwise.} \end{cases} \tag{5.10}$$

Now we state the main result of the section.

5.3. Theorem. *(a) For any choice of the inverse roots $\alpha(q) \neq 0$ $(q \in S)$, and for any s from the convergency region of the series (5.1) there exists a distribution $\mu_s = \mu_{s,\alpha}$ on $\mathbf{Z}_S^\times$ whose values on open compact subsets of the type $a + (M) \subset \mathbf{Z}_S^\times$ are given by the following*

$$\mu_s(a + (M)) = \frac{M^{s-1}}{\alpha(M)} \sum_{S(n)\subset S} B^S(n)P_s(\frac{an}{M})n^{-s}, \tag{5.11}$$

so that the sum in (5.11) is finite and the numbers $B^S(n)$ are defined by (5.10).

(b) For any primitive Dirichlet character χ viewed as a function $\chi : \mathbf{Z}_S^\times \to \mathbf{C}^\times$ the following equality holds

$$\begin{aligned} \int_{\mathbf{Z}_S^\times} \chi\, d\mu_s = & \prod_{q\in S\setminus S(\chi)} (1 - \chi(q)\alpha(q)^{-1}q^{-s})H_q(\overline{\chi}(q)q^{-s})\times \\ & \times \frac{C_\chi^{s-1}}{\alpha(C_\chi)} G(\chi)\mathcal{D}(s,\overline{\chi}), \end{aligned} \tag{5.12}$$

with

$$G(\chi) = \sum_{a \bmod C_\chi} \chi(a)e(\frac{a}{C_\chi})$$

being the Gauss sum, C_χ the conductor, and $S(\chi)$ the support of the conductor of χ.

5.4. The following proof of this theorem differs from that given in [Pan5] and is based on the compatibility criterion (3.4). We check that the sum

$$\sum_{\substack{a \bmod M \\ (a,S)=1}} \chi(a)\mu_s(a + (M)), \tag{5.13}$$

does not depend on the choice of a positive integer M with the condition $C_\chi | M$, $S(M) = S$. This will be provided by a calculation which also implies that (5.13) coincides with the right hand side of (5.12) (and therefore is independent of M).

5.5. Lemma. *For an arbitrary positive integer n and $C_\chi|M$ put*

$$G_{n,M} = \sum_{\substack{a \bmod M \\ (a,S)=1}} \chi(a)e(an/M).$$

Then the following holds

$$G_{n,M}(\chi) = \frac{M}{C_\chi} G(\chi) \sum_{d|(M/C_\chi)} \mu(d)d^{-1}\chi(d)\delta\left(\frac{dn}{(M/C_\chi)}\right)\overline{\chi}\left(\frac{dn}{(M/C_\chi)}\right),$$

in which μ denotes the Möbius function, $\delta(x) = 1$ or 0 according as $x \in \mathbf{Z}$ or not, and we assume that the character χ is primitive modulo C_χ.

The proof of the lemma is deduced from the well known property of the Möbius function:

$$\sum_{d|n} \mu(d) = \begin{cases} 1, & \text{if } n = 1; \\ 0, & \text{if } n > 1. \end{cases}$$

Consequently, $G_{n,M}$ takes the following form:

$$\sum_{\substack{d|(a,M) \\ a \bmod M}} \mu(d)\chi(a)e(an/M) =$$

$$\sum_{d|M} \mu(d) \sum_{a_1 \bmod M/d} \chi(da_1)e(da_1n/M) =$$

$$\sum_{d|M} \mu(d)d^{-1}\chi(d) \sum_{a_1 \bmod M} \chi(a_1)e(da_1n/M) =$$

$$\sum_{d|(M/C_\chi)} \mu(d)d^{-1}\chi(d)\delta\left(\frac{dn}{(M/C_\chi)}\right) \sum_{a_1 \bmod M} \chi(a_1)e\left(da_1n, M,\right),$$

since $\chi(a_1)$ depend only on $a_1 \bmod C_\chi$, and

$$e(a_1dn/M) = e((a_1/C_\chi)(dn/(M/C_\chi))).$$

In the above equality we changed the order of summation, then we replaced the index of summation a by da_1 and extended a system of residue classes $a_1 \bmod M/d$ to a system $a_1 \bmod M$. Now we transform the summation into that one modulo C_χ. It remains to use the well known property of Gauss sums (see, for example, [Shi1], lemma 3.63):

$$G_{n,C_\chi}(\chi) = \overline{\chi}(n)\, G(\chi),$$

establishing the lemma.

5.6. In order to deduce the theorem, we now transform (5.13), taking into account the definition (5.11) and lemma 5.5:

$$\frac{M^{s-1}}{\alpha(M)} \sum_{\substack{a \bmod M \\ (a,M)=1}} \chi(a) \sum_n B^S(n) n^{-s} \sum_{n_1} a_{n_1} e(\frac{ann_1}{M}) n_1^{-s} =$$
$$\frac{M^{s-1}}{\alpha(M)} \sum_n \sum_{n_1} B^S(n) n^{-s} a_{n_1} n_1^{-s} G_{nn_1,M}(\chi) =$$
$$\frac{M^s}{\alpha(M) C_\chi} G(\chi) \sum_n \sum_{n_1} B^S(n) n^{-s} a_{n_1} n_1^{-s} \times \tag{5.14}$$
$$\times \sum_{d|(M/C_\chi)} \frac{\mu(d)}{d} \chi(d)\, \delta\left(\frac{dnn_1}{(M/C_\chi)}\right) \overline{\chi}\left(\frac{dnn_1}{(M/C_\chi)}\right).$$

From the last formula we see that non-vanishing terms in the sum over n and n_1 must satisfy the condition $(M/C_\chi d)|nn_1$. Let us now split n_1 into two factors $n_1 = n_1' \cdot n_1''$ so that $S(n_1') \subset S$, and $(n_1'', S) = 1$. Then

$$\overline{\chi}\left(\frac{dnn_1}{(M/C_\chi)}\right) = \overline{\chi}(n_1'')\overline{\chi}\left(\frac{nn_1'}{(M/C_\chi d)}\right) \tag{5.15}$$

and

$$\delta\left(\frac{dnn_1}{(M/C_\chi)}\right) = \delta\left(\frac{nn_1'}{(M/C_\chi d)}\right), \tag{5.15a}$$

since $(n_1'', M) = 1$. According to (5.2) one has

$$\sum_{S(n_1')\subset S} a_{n_1'} n_1'^{-s} = \prod_{q\in S} F_q(q^{-s})^{-1}. \tag{5.16}$$

Now we use the definition of the finite Euler product (5.9) and of the polynomials $H_q(X)$ which we rewrite here in the form

$$\sum_n B^S(n) n^{-s} \prod_{q\in S} F_q(q^{-s})^{-1} = \prod_{q\in S} (1 - \alpha(q) q^{-s})^{-1}.$$

Consequently,

$$\sum_n B^S(n) n^{-s} \sum_{S(n_1')\subset S} a_{n_1'} n_1'^{-s} = \sum_{S(n_2)\subset S} a_{n_2} n_2^{-s}$$

and for $S(n_2) \subset S$ we have that

$$\alpha(n_2) = \sum_{n_2 = n\cdot n_1'} B^S(n) a_{n_1'}. \tag{5.17}$$

Keeping in mind (5.15) and (5.15a) we transform (5.14) to the following

$$\frac{M^s}{\alpha(M)C_\chi}G(\chi)\sum_n \sum_{d|(M/C_\chi)} \frac{\mu(d)}{d}\chi(d) \sum_{(n_1'',S)=1} \overline{\chi}(n_1'')a_{n_1''}n_1''^{-s}\times$$
$$\times\sum_{n,n_1'} B^S(n)a_{n_1'}(nn_1')^{-s}\delta\left(\frac{nn_1'}{(M/C_\chi d)}\right)\overline{\chi}\left(\frac{nn_1'}{(M/C_\chi d)}\right). \tag{5.18}$$

Now we transform (5.18) with help of the relation (5.17), taking into account that non-zero summands can only occur for such $n_2 = nn_1'$ which are divisible by $M/(C_\chi d)$, i.d. we put $n_2 = (M/C_\chi d)n_3$, $S(n_3) \subset S$. We also note that by the definition of our Dirichlet series we have

$$\sum_{n,n_1'} B^S(n)a_{n_1'}(nn_1')^{-s} = \mathcal{D}(s,\overline{\chi}) \prod_{q\in S\setminus S(\chi)} F_q(\overline{\chi}(q)q^{-s}).$$

Therefore (5.18) transforms to the following:

$$\frac{M^s}{\alpha(M)C_\chi}G(\chi) \prod_{q\in S\setminus S(\chi)} F_q(\overline{\chi}(q)q^{-s})\times$$
$$\sum_{d|(M/C_\chi)} \frac{\mu(d)}{d}\chi(d) \sum_{S(n_3)\subset S} \overline{\chi}(n_3)\alpha\left(\frac{n_3M}{C_\chi d}\right)\left(\frac{n_3M}{C_\chi d}\right)^{-s} =$$
$$\frac{C_\chi^{s-1}}{\alpha(C_\chi)}G(\chi)\mathcal{D}(s,\overline{\chi}) \prod_{q\in S\setminus S(\chi)} F_q(\overline{\chi}(q)q^{-s})\times$$
$$\times \sum_{d|(M/C_\chi)} \mu(d)\, d^{s-1}\chi(d)\alpha(d)^{-1} \sum_{S(n_3)\subset S} \overline{\chi}(n_3)\alpha(n_3)n_3^{-s}.$$

The proof of the theorem is accomplished by noting that

$$\sum_{d|(M/C_\chi)} \mu(d)\, d^{s-1}\chi(d)\alpha(d)^{-1} = \prod_{q\in S\setminus S(\chi)} (1-\chi(q)\alpha(q)^{-1}q^{s-1}),$$
$$\sum_{S(n_3)\subset S} \overline{\chi}(n_3)\alpha(n_3)n_3^{-s} = \prod_{q\in S\setminus S(\chi)} (1-\overline{\chi}(q)\alpha(q)q^{-s})^{-1} =$$
$$\prod_{q\in S\setminus S(\chi)} F_q(\overline{\chi}(q)q^{-s})^{-1}H_q(\overline{\chi}(q)q^{-s}).$$

5.7. Concluding remarks. This construction admits a generalization [Pa7] to the case of rather general Euler products over prime ideals in algebraic number fields. These Euler products have the form

$$\mathcal{D}(s) = \sum_{\mathfrak{n}} a_{\mathfrak{n}}\mathcal{N}(\mathfrak{n})^{-s} = \prod_{\mathfrak{p}} F_{\mathfrak{p}}(\mathcal{N}(\mathfrak{p})^{-s})^{-1},$$

where $\mathfrak{n}$ runs over the set of integral ideals, and $\mathfrak{p}$ over the set of prime ideals of the ring of integers $\mathcal{O}_K$ of a number field K, with $\mathcal{N}(\mathfrak{n})$ denoting the absolute norm of an ideal $\mathfrak{n}$, and $F_\mathfrak{p} \in \mathbf{C}[X]$ being polynomials with the condition $F_\mathfrak{p}(0) = 1$. In [Pa7] we constructed certain canonical distributions, which provide integral representations for special values of Dirichlet series of the type

$$\mathcal{D}(s,\chi) = \sum_{\mathfrak{n}} \chi(\mathfrak{n}) a_\mathfrak{n} \mathcal{N}(\mathfrak{n})^{-s},$$

where χ denote a Hecke character of finite order, whose conductor consists only of prime ideals beloning to a fixed finite set S of non-Archimedean places of K. The main result of [Pa7] provides a generalization of theorem 4.2 of the ealier work of Yu.I.Manin [Man6].

In the construction of non-Archimedean convolutions of Hilbert modular forms given below in chapter 4, we give another approach to local distributions, which differs from the given above and is applicable only to certain Dirichlet series (namely, to convolutions of Rankin type). However, there is an interesting link between these two approaches, which is based on a general construction of p-adic distributions attached to motives. It turns out that both types of distributions can be obtained using a certain transform in the distibution space (see [Co-PeRi], [Co-Schn]).

Chapter 2. Siegel modular forms and the holomorphic projection operator

This chapter contains some preparatory facts which we shall use in the construction of non-Archimedean standard zeta-functions in the next chapter. We recall main properties of Siegel modular forms and of the action of the Hecke algebra on them, as well as the definitions of spinor zeta functions and standard zeta functions (§1), see also [An2], [An7]. Then in §2 we present some standard results on theta series with a Dirichlet character [An-M1], [An-M2], [St2] and recall the definitions of Siegel – Eisenstein series, and of Rankin type convolutions of Siegel modular forms. Also, we recall their relation with the standard zeta functions (Andrianov's identity). In §3 we give an exposition of some recent results of Shimura and P.Feit on real analytic Siegel – Eisenstein series and their analytic continuation in terms of confluent hypergeometric functions [Fe], [Shi7], [Shi9]. These results extend previous results of V.L.Kalinin [K1] and Langlands [Ll1]. In the final §4 a detailed study of holomorphic projection operator and its basic properties is given. The formula of theorem 4.2 provides an explicit formula for computing the holomorphic projection onto the space of holomorphic (not necessarily cusp) modular forms for functions belonging to a wide class of C^∞-Siegel modular forms. Ealier the holomorphic projection operator onto the space of cusp form was studied by J.Sturm [St1], [St2], B.Gross and D.Zagier [Gr-Z] under some restrictive assumptions on the growth of modular forms. Theorem 4.6 gives an explicit description of the action of this operator in terms of Fourier expansions. Here we also establish a very explicit formula (3.36) for the special (critical) values of the confluent hypergeometric function.

Notations

Let A be a commutative ring with identity, then $\mathrm{M}_{r,s}(A)$ denote the set of all $r \times s$ -matrices with coefficients in A. For $z \in \mathrm{M}_r(\mathbf{C})$ put $e_r(z) = e(\mathrm{tr}(z))$ with $e(u) = \exp(2\pi i u)$ for $u \in \mathbf{C}$. We denote by ${}^tz \in \mathrm{M}_{s,r}$ the matrix, which is transpose to $z \in \mathrm{M}_{r,s}(A)$, and write $\xi[\eta]$ for ${}^t\eta\xi\eta$. For a non-degenerate square matrix ξ we put $\xi^* = {}^t\xi^{-1}$. If ξ is a hermitian matrix then we write $\xi \geq 0$ or $\xi > 0$ according as ξ is non negative or positive definite.

Let $\mathfrak{H}_m$ denote the Siegel upper half plane on the degree m,

$$\mathfrak{H}_m = \{z \in \mathrm{M}_m(\mathbf{C}) | {}^tz = z = x + iy, \quad y > 0\},$$

so that $\mathfrak{H}_m$ is a complex analytic variety whose dimension is denoted by $\langle m \rangle = m(m+1)/2$.

Let the symbol A_m denote the lattice of all half integral symmetric matrices in the vector space $V = \{y \in \mathrm{M}_m(\mathbf{R}) | {}^ty = y\}$. This lattice is dual to the lattice $L = \mathrm{M}_m(\mathbf{Z}) \cap V$ with respect to the pairing given by $(u, v) \mapsto e_m(uv)$. For a function $f : \mathfrak{H}_m \to \mathbf{C}$ of the form

$$f = \sum_{\xi \in A_m} c(\xi) e_m(\xi z) \quad (z \in \mathfrak{H}_m)$$

and for a positive integer A we use the notations

$$f|V(A)(z) = f(Az) = \sum_{\xi \in A_m} c(\xi) e_m(A\xi z),$$

$$f|U(A)(z) = \sum_{\xi \in A_m} c(A\xi) e_m(\xi z),$$

$$f^\rho = \sum_{\xi \in A_m} \overline{c(\xi)} e_m(\xi z)$$

as well as the notation of A.N.Andrianov for the action of the Frobenius elements $\Pi^+(q)$, $\Pi^-(q)$ given in 1.8. Moreover, for $A \geq 1$ and an integer k we put

$$f|W(A)(z) = f|_k W(A)(z) = \det(\sqrt{A} z)^k f(-(Az)^{-1}),$$

so that

$$(f|W(A))|W(A) = (-1)^{mk} f.$$

§1. Siegel modular forms and Hecke operators

1.1. Symplectic group and Siegel upper half plane. (sèe [An2], [An7], [Shi4], [Sie3], [Fr], [Maa]). Let $G = \mathrm{GSp}$ be the algebraic subgroup of GL_{2m} defined by

$$G_A = \{\alpha \in \mathrm{GL}_{2m}(A) \mid {}^t\alpha J_m \alpha = \nu(\alpha) J_m, \nu(\alpha) \in A^\times\}, \tag{1.1}$$

for any commutative ring A, where

$$J_m = \begin{pmatrix} 0_m & -1_m \\ 1_m & 0_m \end{pmatrix}.$$

The elements of G_A are characterized by the conditions

$$b^t a - a^t b = d^t c - c^t d = 0_m, \quad d^t a - c^t b = 1_m, \tag{1.2}$$

and if

$$\alpha = \begin{pmatrix} a & b \\ c & d \end{pmatrix} \in G_A \text{ then } \alpha^{-1} = \nu(\alpha)^{-1} \begin{pmatrix} {}^t d & -{}^t b \\ -{}^t c & {}^t a \end{pmatrix}.$$

The multiplier ν defines a homomorphism $\nu : G_A \to A^\times$ so that $\nu(\alpha)^{2m} = \det(\alpha)^2$ and $\mathrm{Ker}\,\nu$ is denoted by $\mathrm{Sp}_m(A)$. We also put

$$G_\infty = G_{\mathbf{R}}, \quad G_{\infty+} = \{\alpha \in G_\infty \mid \nu(\alpha) > 0\}, \quad G_{\mathbf{Q}+} = G_{\infty+} \cap G_{\mathbf{Q}}. \tag{1.3}$$

The group $G_{\infty+}$ acts transitively on the upper half plane $\mathfrak{H}_m$ by the rule

$$z \to \alpha(z) = (az+b)(cz+d)^{-1} \quad \left(\alpha = \begin{pmatrix} a & b \\ c & d \end{pmatrix} \in G_{\infty+}, z \in \mathfrak{H}_m\right)$$

so that scalar matrices act trivially, and $\mathfrak{H}_m$ can be identified with a homogeneous space of the group $\mathrm{Sp}_m(\mathbf{R})$. Let K_m denote the stabilizer of the point $i1_m \in \mathfrak{H}_m$ in the group $\mathrm{Sp}_m(\mathbf{R})$,

$$K_m = \{\alpha \in \mathrm{Sp}_m(\mathbf{R}) \mid \alpha(i1_m) = i1_m\},$$

then there is a bijection $\mathrm{Sp}_m(\mathbf{R})/K_m \simeq \mathfrak{H}_m$ and $K_m = \mathrm{Sp}_m(\mathbf{R}) \cap \mathrm{SO}_{2m}$. The group K_m is a maximal compact subgroup of the Lie group $\mathrm{Sp}_m(\mathbf{R})$ and it can be identified with the group $\mathrm{U}(m)$ of all unitary $m \times m$-matrices via the map $\alpha = \begin{pmatrix} a & b \\ c & d \end{pmatrix} \mapsto a+ib$. We adopt also notations

$$\begin{gathered} dx = \prod_{i \le j} dx_{ij}, \quad dy = \prod_{i \le j} dy_{ij}, \quad dz = dx\,dy, \\ d^\times y = \det(y)^{-(m+1)/2} dy, \quad d^\times z = \det(y)^{-(m+1)} dz, \end{gathered} \tag{1.4}$$

where $z = x + iy, \ x = (x_{ij}) = {}^t x, \ y = (y_{ij}) = {}^t y > 0$. Then $d^\times z$ is a differential on $\mathfrak{H}_m$ invariant under the action of the group $G_{\infty+}$, and the measure $d^\times y$ is invariant under the action of elements $a \in \mathrm{GL}_m(\mathbf{R})$ on

$$Y = \{y \in \mathrm{M}_m(\mathbf{R}) \mid {}^t y = y > 0\}$$

defined by the rule $y \mapsto {}^t a y a$.

1.2. Siegel modular forms. Let us consider Siegel modular group $\Gamma^m = \mathrm{Sp}_m(\mathbf{Z})$ and let $\Gamma \subset G_{\mathbf{Q}+}$ be an arbitrary congruence subgroup. This means that Γ is commeasurable with Γ^m in $G_{\mathbf{Q}+}$ modulo its center (i.e. as a group of transformations of $\mathfrak{H}_m$) and $\Gamma \supset \Gamma^m(N)$ for some $N \in \mathbf{N}$,where

$$\Gamma^m(N) = \left\{\gamma = \begin{pmatrix} a & b \\ c & d \end{pmatrix} \in \Gamma^m \mid \gamma \equiv 1_{2m} \,(\mathrm{mod}\, N)\right\}$$

is the main congruence subgroup of level N in Γ^m. In order to give the general definition of modular forms we consider a rational representation $\rho : \mathrm{GL}_m(\mathbf{C}) \to \mathrm{GL}_r(\mathbf{C})$ which will also be denoted by ρ. For $\alpha = \begin{pmatrix} a & b \\ c & d \end{pmatrix} \in G_{\infty+}$ and for any complex valued function $f : \mathfrak{H}_m \to \mathbf{C}^r$ we use the notation

$$f \mid_\rho \alpha(z) = \rho(cz+d)^{-1} f(\alpha(z)). \tag{1.5}$$

Definition. A function $f : \mathfrak{H}_m \to \mathbf{C}^r$ is called a *holomorphic modular form* of weight ρ on Γ if the following conditions (1.6) – (1.8) are satisfied:

$$f \mid_\rho = f, \tag{1.6}$$

$$f \text{ is holomorphic on } \mathfrak{H}_m, \tag{1.7}$$

$$\text{if } m = 1, \text{ then } f \text{ is holomorphic at cusps of } \Gamma. \tag{1.8}$$

Let $\mathcal{M}_\rho(\Gamma)$ be the complex vector space of functions satisfying the above conditions. For each $f \in \mathcal{M}_\rho(\Gamma)$ there is the following Fourier expansion

$$f(z) = \sum_\xi c(\xi) e_m(\xi z),$$

where $c(\xi) \in \mathbf{C}^r, \xi$ run over all $\xi = {}^t\xi \in \mathrm{M}_m(\mathbf{Q}), \xi \geq 0$ (for $m > 1$ the last condition automatically follows by the Koecher principle). More precisely, let M be the smallest integer such that

$$\Gamma \supset \left\{\begin{pmatrix} 1_m & Mu \\ 0 & 1_m \end{pmatrix} \mid u \in \mathrm{M}_m(\mathbf{Z}),\ \ {}^tu = u\right\}$$

and we put

$$A = A_m = \{\xi = (\xi_{ij}) \in \mathrm{M}_m(\mathbf{R}) \mid \xi = {}^t\xi,\ \xi_{ij}, 2\xi_{ii} \in \mathbf{Z}\},$$
$$B = B_m = \{\xi \in A \mid \xi \geq 0\}, C = C_m = \{\xi \in A \mid \xi > 0\}$$

Then A_m is a lattice in the $\mathbf{R}$-vector space of symmetric matrices $V = \{x \in \mathrm{M}_m(\mathbf{R}) \mid {}^tx = x\}$ dual to the lattice $L = \mathrm{M}_m(\mathbf{Z}) \cap V$ with respect to the action $(\xi, x) \mapsto e_m(\xi x)$ and for each $f \in \mathcal{M}_\rho(\Gamma)$ there is the following Fourier expansion

$$f(z) = \sum_{\xi \in M^{-1}B} c(\xi) e_m(\xi z), \tag{1.9}$$

Moreover, for each $\sigma \in G_{\mathbf{Q}+}$ we have that $f\mid_\rho \sigma \in \mathcal{M}_\rho(\Gamma(\sigma))$,where $\Gamma(\sigma)$ is a congruence subgroup,

$$(f|_\rho\sigma)(z) = \sum_{\xi \in M_\sigma^{-1}B} c_\sigma(\xi)e_m(\xi z), \tag{1.10}$$

with $c_\sigma(\xi) \in \mathbf{C}^r, M_\sigma \in \mathbf{N}$. A form f is called a cusp form if for all ξ with $\det(\xi) = 0$ in expansion (1.10) one has $c_\sigma(\xi) = 0$ for all $\sigma \in G_{\mathbf{Q}+}$ that is

$$(f|_\rho\sigma)(z) = \sum_{\xi \in M_\sigma^{-1}C} c_\sigma(\xi)e_m(\xi z),$$

We denote by $\mathcal{S}_\rho(\Gamma) \subset \mathcal{M}_\rho(\Gamma)$ the subspace of cusp forms.

Definition of the vector spaces $\mathcal{M}(N,\psi)$. Let us consider congruence subgroups $\Gamma_1^m(N) \subset \Gamma_0^m(N) \subset \Gamma^m(N) = \mathrm{Sp}_m(\mathbf{Z})$, defined by

$$\Gamma_0^m(N) = \{\gamma = \begin{pmatrix} a & b \\ c & d \end{pmatrix} \in \mathrm{Sp}_m(\mathbf{Z}) \mid c \equiv 0_m(\mathrm{mod} N)\},$$
$$\Gamma_1^m(N) =$$
$$\{\gamma = \begin{pmatrix} a & b \\ c & d \end{pmatrix} \in \mathrm{Sp}_m(\mathbf{Z}) \mid c \equiv 0_m(\mathrm{mod} N), \det(a) \equiv 1(\mathrm{mod} N)\},$$

and let $r = 1$, $\rho(x) = \rho_k(x) = \det(x)^k$ $(k \in \mathbf{N})$. Then the vector space $\mathcal{M}(\Gamma_1^m(N))$ has already been defined, and we set

$$\mathcal{M}_m^k(N,\psi) = \{f \in \mathcal{M}_\rho(\Gamma_1^m(N)) \mid f\mid_\rho \gamma = \psi(\det(a))f \text{ for all } \gamma = \begin{pmatrix} a\, b \\ c\, d \end{pmatrix} \in \Gamma_0^m(N)\}, \tag{1.11}$$

where ψ is a Dirichlet character modulo N. Elements $f \in \mathcal{M}_m^k(N,\psi)$ admit a Fourier expansion of the form

$$f(z) = \sum_{\xi \in B_m} c(\xi)e_m(\xi z), \tag{1.12}$$

with $z \in \mathfrak{H}_m, c(\xi) \in \mathbf{C}$, and the condition

$$c(u\xi{}^tu) = \psi(\det u)\det u^k c(\xi) \quad (\xi \in B_m, u \in \mathrm{GL}_m(\mathbf{Z})). \tag{1.12a}$$

Put

$$\mathcal{S}_m^k(N,\psi) = \mathcal{M}_m^k(N,\psi) \cap \mathcal{S}_{\rho_k}(\Gamma_1^k(N)).$$

The Petersson scalar product. For$f \in \mathcal{S}_m^k(N,\psi)$ and $h \in \mathcal{M}_m^k(N,\psi)$ the Petersson scalar product is defined by

$$\langle f, h\rangle_N = \int_{\Phi_0(N)} \overline{f(z)}h(z)\det y^k d^\times z, \tag{1.13}$$

where $\Phi_0(N) = \Gamma_0^m(N)\backslash\mathfrak{H}_m$ is a fundamental domain for the group $\Gamma_0^m(N)$.

The Siegel operator connects the vector spaces $\mathcal{M}_m^k(N,\psi)$ for different values of m. If $f \in c$, $z \in \mathfrak{H}_{m-1}$ and $\lambda > 0$ then we have that

$$\begin{pmatrix} z' & 0 \\ 0 & i\lambda \end{pmatrix} \in \mathfrak{H}_m,$$

and it follows from (1.12) that there exists the limit

$$(\Phi f)(z\prime) = \lim_{\lambda\to\infty} f\left(\begin{pmatrix} z' & 0 \\ 0 & i\lambda \end{pmatrix}\right) = \sum_{\xi'\in B_{m-1}} c\left(\begin{pmatrix} \xi' & 0 \\ 0 & 0 \end{pmatrix}\right) e_{m-1}(\xi' x'), \tag{1.14}$$

where $c(\xi)$ are the Fourier coefficients of f. Then

$$\Phi f \in \mathcal{M}_{m-1}^k(N,\psi),$$

(we put $\mathcal{M}_0^k(N,\psi) = 0$). We then have that

$$\mathcal{S}_m^k(N,\psi) \subseteq \mathrm{Ker}\Phi = \{f \in \mathcal{M}_m^k(N,\psi) \mid \Phi f = 0\}$$

(for $N = 1$ both sets coincide, see[Maa]).

Estimates for Fourier coefficients. If $f \in \mathcal{S}_m^k(N,\psi)$ then there is the following upper estimate

$$\mid f(z) \mid = \mathcal{O}((\det y)^{-k/2}) \quad (z = x + iy \in \mathfrak{H}_m), \tag{1.15}$$

which provides us also with the estimate

$$\mid c(\xi) \mid = \mathcal{O}(\det(\xi)^{k/2}) \tag{1.15a}$$

For modular (not necessary cusp) forms

$$f(z) = \sum_{\xi\in B_m} c(\xi) e_m(\xi z) \in \mathcal{M}_m^k(N,\psi)$$

there is the upper estimate of their growth:

$$\mid c(\xi) \mid \le c_1 \prod_{j=1}^m (1 + \lambda_j^k), \tag{1.16}$$

with $\lambda_1, \cdots, \lambda_m$ being eigenvalues of the matrix $y, z = x + iy$ (see [St2], p.335). In this situation one has also the following estimate

$$\mid c(\xi) \mid \le c_2(\det(\xi')^k) \tag{1.16a}$$

in which c_2 is a positive constant depending only on f, $\xi = {}^t u \begin{pmatrix} \xi' & 0 \\ 0 & 0 \end{pmatrix} u$, $u \in \mathrm{SL}_m(\mathbf{Z})$, $\xi' \in B_r$, $\det \xi' > 0$, $r \le m$.

We refer the reader to [Fo1], [Ki1], [Rag1], [Rag4] for a more detailed discussion of various estimates for Fourier coefficients and for growth of Siegel modular forms, as well as for some interesting applications to quadratic forms.

1.3. The Hecke algebra (see [An7], [Bö4], [Sat]). Let q be a prime, $q \not| N$,

$$\Delta = \Delta_q^m =$$
$$\{\alpha = \begin{pmatrix} a & b \\ c & d \end{pmatrix} \in G_{\mathbf{Q}+} \cap \mathrm{GL}_{2m}(\mathbf{Z}[q^{-1}]) | \nu(\alpha)^{\pm} \in \mathbf{Z}[q^{-1}], c \equiv 0_m \,(\mathrm{mod} N)\}$$

be a subgroup in $G_{\mathbf{Q}+}$ containing $\Gamma = \Gamma_0^m(N)$. The Hecke algebra

$$\mathcal{L} = \mathcal{L}_q^m(N) = \mathcal{D}_{\mathbf{Q}}(\Gamma, \Delta)$$

over $\mathbf{Q}$ is then defined as a $\mathbf{Q}$-linear space generated by the double cosets $(g) = (\Gamma g \Gamma)$, $g \in \Delta$ of the group S with respect to the subgroup Γ, for which multiplication is defined by the standard rule (see [An7], [Shi1] and in 1.7 below). We recall the description of the structure of $\mathcal{L} = \mathcal{L}_q^m(N)$, $(q \not| N)$: for each $j, 1 \le j \le m$ let us denote by w_j an automorphism of the algebra $\mathbf{Q}[x_0^{\pm 1}, x_1^{\pm 1}, \cdots, x_m^{\pm 1}]$ defined on its generators by the rule:

$$x_0 \mapsto x_0 x_j, \; x_j \mapsto x_j^{-1}, x_i \mapsto x_i \quad (1 \le i \le m, i \ne j).$$

Then the automorphisms w_j and the permutation group Σ_m of the variables x_i $(1 \le i \le m)$ generate together the Weyl group $W = W_m$, and there is the Satake isomorphism:

$$Sat : \mathcal{L} \xrightarrow{\sim} \mathbf{Q}[x_0^{\pm 1}, x_1^{\pm 1}, \cdots, x_m^{\pm 1}]^{W_m} \tag{1.17}$$

where W_m indicates the subalgebra of elements fixed by W_m. For any commutative $\mathbf{Q}$-algebra A the group W_m acts on the set $(A^\times)^{m+1}$, therefore any homomorphism of $\mathbf{Q}$-algebras $\lambda : \mathcal{L} \to A$ can be identified with some element

$$(\alpha_0, \alpha_1, \cdots, \alpha_m) \in [(A^\times)^{m+1}]^{W_m}. \tag{1.18}$$

An explicit description of the Satake isomorphism is given below in 1.7.

1.4. Any double coset $(g) = (\Gamma g \Gamma)$ $(g \in \Delta = \Delta_q^m(N))$ can be represented is a disjoint union of left cosets:

$$(g) = \cup_{i=1}^{t(g)} \Gamma g_i,$$

therefore any element $X \in \mathcal{L}$ of the Hecke algebra $\mathcal{L}$ takes the form of a finite linear combination

$$X = \sum_{i=1}^{t(X)} v_i(\Gamma g_i),$$

with $v_i \in \mathbf{Q}, g_i \in \Delta$. In order to define the Hecke operators we put for any $g = \begin{pmatrix} a & b \\ c & d \end{pmatrix} \in \Delta$

$$(f|_{k,\psi} g)(z) = (\det g)^{k-(m+1)/2} \psi(\det a) \det(cz+d)^{-k} f(g(z)) \tag{1.19}$$

(this convenient notation, suggested by Petersson and Andrianov, is especially useful when dealing with Hecke operators in their normalized form, compare with (1.5)). With this notation the automorphy condition can be rewritten as follows

$$(f|_{k,\psi}\gamma)(z) = f \quad \text{for all } \gamma \in \Gamma = \Gamma_0^m(N). \tag{1.20}$$

In this case for any

$$X = \sum_{i=1}^{t(X)} v_i(\Gamma g_i) \in \mathcal{L}$$

we have that the expression

$$f|X = \sum_{i=1}^{t(X)} v_i f|_{k,\psi}\, g_i, \tag{1.21}$$

is well defined and $f|X \in \mathcal{M}_k^m(N,\psi)$ so that the formula (1.21) gives a representation of the Hecke algebra $\mathcal{L} = \mathcal{L}_q^m(N)$ on the complex vector space $\mathcal{M}_k^m(N,\psi)$ $(q \not| N)$.

1.5. Hecke polynomials. Following A.N.Andrianov[An7], let us consider polynomials

$$\begin{aligned} \tilde{Q} &= \tilde{Q}(x_0, x_1, \cdots, x_m; z) = \\ &= (1 - x_0 z) \prod_{r=1}^{m} \prod_{1 \le i_1 < \cdots < i_r \le m} (1 - x_0 x_{i_1} \cdots x_{i_r} z), \end{aligned} \tag{1.22}$$

$$\tilde{R}(z) = \prod_{i=1}^{m} (1 - x_i^{-1} z)(1 - x_i z) \in \mathbf{Q}[x_0^{\pm 1}, \cdots, x_m^{\pm 1}][z]. \tag{1.23}$$

From the definition it follows that the coefficients of the powers of the variable z all belong to the subring

$$\mathbf{Q}[x_0^{\pm 1}, x_1^{\pm 1}, \cdots, x_m^{\pm 1}]^{W_m}.$$

Therefore by Satake isomorphism (1.17) we have that there exist polynomials

$$Q(z) = \sum_{i=0}^{2^m} (-1)^i T_i z^i, \quad R(z) = \sum_{i=0}^{2m} (-1)^i R_i z^i \in \mathcal{L}[z] \tag{1.24}$$

over the associative commutative ring $\mathcal{L} = \mathcal{L}_q^m(N)$ such that

$$\tilde{Q}(z) = \sum_{i=0}^{2^m} (-1)^i \tilde{T}_i z^i, \quad \tilde{R}(z) = \sum_{i=0}^{2m} (-1)^i \tilde{R}_i z^i$$

with $\tilde{X} = Sat\, X$, $X \in \mathcal{L}$. The polynomials ${\Delta'_M}^{\pm 1}, R_i (1 \le i \le m-1)$ and T_1 with

$$\tilde{\Delta}'^{\pm 1}_M = x_0^2 x_1 \cdots x_m, \quad R_i = S_i(x_1, \cdots x_m; x_1^{-1}, \cdots x_m^{-1}),$$

$$\tilde{T}_1 = x_0 \sum_{i=1}^{m} S_i(x_1, \cdots x_m) = x_0 \prod_{i=1}^{m} (1 + x_i),$$

can be taken as generators of the Hecke algebra, where S_i denotes the elementary symmetric polynomial of degree i of the corresponding set of variables.

1.6. The spinor zeta function and the standard zeta function. Let $f \in \mathcal{M}_m^k(N,\psi)$ be an eigenfunction of all Hecke operators $f \to f|X, X \in \mathcal{L}_q^m(N)$ with q being prime numbers, $q \not| N$, so that $f|X = \lambda_f(X)f$. Then the numbers $\lambda_f(X) \in \mathbf{C}$ define a homomorphism $\lambda_f : \mathcal{L} \to \mathbf{C}$ (see 1.3), which is uniquely determined by a $(m+1)$-tuple of the numbers

$$(\alpha_0, \alpha_1, \cdots \alpha_m) = (\alpha_{0,f}(q), \alpha_{1,f}(q), \cdots \alpha_{m,f}(q)) \in [(\mathbf{C}^\times)^{m+1}]^{W_m} \tag{1.25}$$

which are called *the Satake q-parameters* of the modular form f.

Now let the variables $x_0, x_1, \cdots, x_m$ in (1.22), (1.23) be equal to the corresponding Satake q-parameters $\alpha_{0,f}(q), \alpha_{1,f}(q), \cdots \alpha_{m,f}(q)$

$$\begin{aligned} Q_{f,q}(z) = & \\ & = (1-\alpha_0 z)\prod_{r=1}^{m} \prod_{1\le i_1<\cdots<i_r\le m} (1-\alpha_0\alpha_{i_1}\cdots\alpha_{i_r} z), \end{aligned} \tag{1.26}$$

$$R_{f,q}(z) = \prod_{i=1}^{m}(1-\alpha_i^{-1}z)(1-\alpha_i z) \in \mathbf{Q}[\alpha_0^{\pm 1}, \cdots, \alpha_m^{\pm 1}]. \tag{1.26a}$$

From the description of the Hecke polynomials in 1.5 it follows then that the coefficients of (1.26) can be expressed in terms of the eigenvalues $\lambda_f(X)$ of the Hecke operators $X = T_i, R_i$ from (1.24). Next we put

$$\begin{aligned} Z^{(q)}(s,f) &= Q_{f,q}(q^{-s})^{-1} = \\ &= \{(1-\alpha_0(q)q^{-s})\prod_{r=1}^{m}\prod_{1\le i_1<\cdots<i_r\le m}(1-\alpha_0(q)\alpha_{i_1}(q)\cdots\alpha_{i_r}(q)q^{-s})\}^{-1}, \end{aligned} \tag{1.27}$$

and define the spinor zeta function $Z(s,f)$ of the modular form $f \in \mathcal{M}_m^k(N,\psi)$ by

$$Z(s,f) = \prod_{q \not| N} Z^{(q)}(s,f) \tag{1.27a}$$

Complex analytic properties of zeta functions $Z(s,f)$ were investigated by A. N. Andrianov [An2] in the case $m=2$. For $m=1$ we have that

$$\alpha_0(q) + \alpha_0(q)\alpha_1(q) = a(q), \quad \alpha_0^2(q)\alpha_1(q) = \psi(q)q^{k-1},$$

where $f(z) = \sum_{n=0}^{\infty} a(n)e(nz)$ is the Fourier expansion of a normalized elliptic cusp Hecke eigenform, so that the zeta function

$$Z(s,f) = \prod_{q\not| N}[1 - a(q)q^{-s} + \psi(q)q^{k-1-2s}]^{-1} = \sum_{\substack{n=1\\(n,N)=1}}^{\infty} a(n)n^{-s}$$

coincides essentially with the Mellin transform

$$L(s,f)=\sum_{n=1}^{\infty}a(n)n^{-s}$$

of f. Notice that all of the inverse roots of the Hecke polynomial $Q_{f,q}(z)$ can be obtained by the action of the Weyl group W_m on $\alpha_0(q)$, and

$$\begin{aligned}&Q_{f,q}(z)=\\&1-\lambda_f(q)q^{-s}+\cdots+q^{2^{m-1}(mk-m(m+1)/2)}z^{2^m}=\\&(1-\alpha_0(q)z)\prod_{r=1}^{m}\prod_{1\le i_1<\cdots<i_r\le m}(1-\alpha_0(q)\alpha_{i_1}(q)\cdots\alpha_{i_r}(q)z),\end{aligned}\tag{1.28}$$

where

$$\lambda_f(q)=\lambda_f(T(q)),T((q))=\sum_{\nu(g)=q}(\Gamma g\Gamma)$$

are the Hecke opertor for the group $\Gamma=\Gamma_0^m(N), {}^tgJ_mg=\nu(g)J_m$.

According to the generalized Ramanujan – Petersson conjecture for a cusp eigenform $f\in\mathcal{S}_m^k(N,\psi)$ the absolute values of all inverse roots of the polynomial (1.28) should be equal to each other (and therefore be equal to $q^{-(2mk-m(m+1)/4)}$). In the case $m=1$ this conjecture is valid; it was deduced by Deligne [De1] from the Weil conjectures, also proven by him [De2]. However, for larger values of m the analogous conjecture is no longer true. Various counterexamples are known, see [Kur],[H-PSh1],[H-PSh2] which for the case $m=2$ reflect the fact that some cusp forms of degree 2 (namely those belonging to the Maass subspace) are lifted from elliptic cusp forms (i.e. of degree 1) via the Saito – Kurokawa lifting [An8], [Koj]. According to the modified Ramanujan – Petersson conjecture these properties of the inverse roots and polynomials $Q_{f,q}$ should hold for those cusp forms, which can not be obtained from the forms of smaller degree by a lifting of the above type.

The standard zeta function $\mathcal{D}(s,f,\chi)$ of $f\in\mathcal{M}_m^k(N,\psi)$ is the product $\mathcal{D}(s,f,\chi)=\prod_{q\nmid N}\mathcal{D}^{(q)}(s,f,\chi)$ with

$$\mathcal{D}^{(q)}(s,f,\chi)=(1-\chi(q)\psi(q)q^{-s})^{-1}R_{f,q}(\chi(q)\psi(q)q^{-s})^{-1},$$

that is

$$\begin{aligned}&\mathcal{D}(s,f,\chi)=\\&=\prod_q\left\{\left(1-\frac{\chi(q)\psi(q)}{q^s}\right)\prod_{i=1}^{m}\left(1-\frac{\chi(q)\psi(q)\alpha_i(q)^{-1}}{q^s}\right)\left(1-\frac{\chi(q)\psi(q)\alpha_i(q)}{q^s}\right)\right\}^{-1},\end{aligned}\tag{1.29}$$

where χ is a Dirichlet character mod M. Analytic properties of the standard zeta functions were investigated by A. N. Andrianov and V. L. Kalinin [An-K] in the case of even degree m, and more recently S. Böcherer [Bö4] extended these results to the case of arbitrary degree using a different approach. For $m = 1$ and a normalized cusp eigenform $f(z) = \sum_{n=0}^{\infty} a(n)e(nz) \in \mathcal{M}_1^k(N, \psi)$ we have that

$$\mathcal{D}(s, f, \chi) = L_{2,f}(s + k - 1, \chi),$$

where

$$L_{2,f}(s, \chi) = L_{NM}(2s - 2k + 2, \chi^2\psi^2) \sum_{n=1}^{\infty} \chi(n)a(n^2)n^{-s} \tag{1.30}$$

is the symmetric square of the modular form f, see [An3], [An-K], [Pa3], [Sch], [Shi2] [Za].

1.7. Non-commutative extensions of the Hecke algebra and the Satake isomorphism. Now we recall that multiplication in the Hecke algebra $\mathcal{L} = \mathcal{D}_{\mathbf{Q}}(\Gamma, \Delta)$ is defined by use of the larger vector space $\mathcal{V} = \mathcal{V}_{\mathbf{Q}}(\Gamma, \Delta)$ over $\mathbf{Q}$ consisting of all $\mathbf{Q}$-linear combinations of left cosets of the form (Γg), $g \in \Delta = \Delta_q^m(N)$. If $\gamma \in \Gamma$ and

$$X = \sum_{i=1}^{t(X)} v_i(\Gamma g_i) \in \mathcal{V}$$

then the formula

$$X \cdot \gamma = \sum_{i=1}^{t(X)} v_i(\Gamma g_i \gamma) \in \mathcal{V}$$

defines a right action of the group Γ on $\mathcal{V}$ so that the algebra $\mathcal{L} = \mathcal{D}_{\mathbf{Q}}(\Gamma, \Delta)$ coincides with the subspace of all elements in $\mathcal{V}_{\mathbf{Q}}(\Gamma, \Delta)$ pointwise fixed by Γ via the inclusion $\mathcal{L} \to \mathcal{V}$ which sends a double cosets

$$(g) = (\Gamma g \Gamma) \ \ (g \in \Delta = \Delta_q^m(N)) = \cup_{i=1}^{t(g)} \Gamma g_i$$

to the formal sum $\sum_{i=1}^{t(g)} \Gamma g_i$. If

$$X = \sum_{i=1}^{t(X)} a_i(\Gamma g_i), \quad Y = \sum_{j=1}^{t(Y)} b_j(\Gamma h_j) \tag{1.32}$$

are two elements in $\mathcal{D}_{\mathbf{Q}}(\Gamma, \Delta) \in \mathcal{V}$ then the element

$$X \cdot \mathcal{V} = \sum_{i,j=1} a_i b_j (\Gamma g_i h_j) \in \mathcal{V}_{\mathbf{Q}}(\Gamma, \Delta)$$

is well defined and also belongs to $\mathcal{D}_{\mathbf{Q}}(\Gamma, \Delta) \in \mathcal{V}$. This construction can be applied to a large variety of couples (Γ, Δ), for which Γ is a subgroup of a (semi)group Δ such that any double coset $(g) = (\Gamma g \Gamma)$ $(g \in \Delta = \Delta_q^m(N))$ can be represented is a disjoint union of left cosets:

$$(g) = \cup_{i=1}^{t(g)} \Gamma g_i$$

(the theory of *Hecke couples*, see [An7], chapter 1). In particular, for

$$\Gamma_0 = \Gamma_0^m = \left\{ g = \begin{pmatrix} a & b \\ c & d \end{pmatrix} \in \Gamma | \ c = 0 \right\},$$

$$\Delta_0 = \Delta_{0,q}^m = \left\{ g = \begin{pmatrix} a & b \\ c & d \end{pmatrix} \in \Delta_{0,q}^m | \ c = 0 \right\}$$

we thus obtain an associative (but no longer commutative) ring

$$\mathcal{L} = \mathcal{L}_{0,q}^m(N) = \mathcal{D}_{\mathbf{Q}}(\Gamma_0, \Delta_0).$$

From the theory of elementary divisors it follows that $\Gamma\Delta_0 = \Delta$; therefore we get the homomorphism $\varepsilon : \mathcal{L} \to \mathcal{L}_0$ defined as follows: for each

$$X = \sum_{i=1}^{t(X)} v_i(\Gamma g_i) \in \mathcal{L} \subset \mathcal{V}$$

we may assume that

$$g_i = \begin{pmatrix} q^{\nu_i} d_i^* & b_i \\ 0 & d_i \end{pmatrix},$$

where $q^{\nu_i} = \nu(g_i)$, and we put

$$\varepsilon_0(X) = \sum_{i=1}^{t(X)} v_i(\Gamma_0 \begin{pmatrix} q^{\nu_i} d_i^* & b_i \\ 0 & d_i \end{pmatrix}) \tag{1.33}$$

Similar arguments show also that the matrix d_i can be chosen in the form

$$d_i = \begin{pmatrix} q_i^\delta(1) & * & \cdots & * \\ 0 & q_i^\delta(2) & \cdots & * \\ \cdots & \cdots & \cdots & \cdots \\ 0 & 0 & \cdots & q_i^\delta(m) \end{pmatrix} \quad (\delta_i(1) \le \cdots \le \delta_i(m)) \tag{1.34}$$

with the exponents $\delta_i(j)$, which are unuquely determined by d_i. Now for

$$X = \sum_{i=1}^{t(X)} v_i(\Gamma g_i) \in \mathcal{L}_0$$

we put by definition

$$\Phi(X) = \sum_{i=1}^{t(X)} v_i x_0^{\nu_i} \prod_{j=1}^{m} (x_j q^{-j}) \delta_i(j). \tag{1.35}$$

The Satake isomorphism is then described as the composition of homomorphisms (1.33) and (1.35):

$$Sat = \Phi\varepsilon_0 : \mathcal{L} \xrightarrow{\sim} \mathbf{Q}[x_0^{\pm 1}, x_1^{\pm 1}, \cdots, x_m^{\pm 1}]^{W_m} \tag{1.36}$$

As a consequence of the construction we get that for an arbitrary homomorphism of algebras $\lambda : \mathcal{L} \to \mathbf{C}$ $(\lambda = \lambda_q, q \not| N, \mathcal{L} = \mathcal{L}_q^m(N)$, attached to the $(m+1)$-tuple of q-parameters

$$(\alpha_0, \alpha_1, \cdots, \alpha_m) \in [(\mathbf{C}^\times)^{m+1}]^{W_m},$$

and for any element

$$X = \sum_{i=1}^{t(X)} v_i(\Gamma g_i) \in \mathcal{L} \ \ with\ g_i = \begin{pmatrix} q^{\nu_i} d_i^* & b_i \\ 0 & d_i \end{pmatrix}$$

and d_i is chosen in the form (1.34) the following equality holds:

$$\lambda(X) = \sum_{i=1}^{t(X)} v_i \alpha_0^{\nu_i} \prod_{j=1}^{m} (\alpha_j q^{-j}) \delta_i(j). \qquad (1.37).$$

If$\lambda = \lambda_{f,q}$ for a modular form $f \in \mathcal{M}_m^k(N, \psi)$ then the numbers $\alpha_0, \alpha_1, \cdots, \alpha_m$ satisfy the following relation

$$\alpha_0^2 \alpha_1 \cdots \alpha_m = \psi(q)^m q^{km - m(m+1)/2}, \qquad (1.38)$$

which follows directly from the formula (1.37) and the definition (1.19) applied to the operator $X = (g)$ with $g = q1_{2m}$ (see[An7]).

1.8. Action of the Hecke operators on Fourier expansions. Let

$$f(z) = \sum_{\xi \in B_m} c(\xi) e_m(\xi z) \in \mathcal{M}_m^k(N, \psi) \quad (\xi \in B_m)$$

and for a fixed prime q

$$X = \sum_{i=1}^{t(X)} v_i(\Gamma_0 g_i) \in \mathcal{L}_0 \quad (\Gamma_0 = \Gamma_0^m, \mathcal{L}_0 = \mathcal{L}_{0,q}^m)$$

be an arbitrary element of the extended Hecke algebra $\mathcal{L}_0$ with

$$g_i = \begin{pmatrix} q^{\nu_i} d_i^* & b_i \\ 0 & d_i \end{pmatrix} \in \Delta_{0,q}^m. \qquad (1.39)$$

Let coefficients $(c|X)(\xi)$ be defined by the equality

$$(f|X)(z) = \sum_{\xi \in B_m} (c|X)(\xi) e_m(\xi z). \qquad (1.40)$$

Then we have that

$$\begin{aligned}(c|X)(\xi) = \sum_{i=1}^{t(X)} v_i q^{\nu_i(mk - \langle m \rangle)} (\det d_i)^{-k} \times \\ \times \psi(q^{m\nu_i} (\det d_i)^{-1}) e_m(q^{-\nu_i}\, {}^t b_i d_i) c(q^{-\nu_i} d_i \xi\, {}^t d_i),\end{aligned} \qquad (1.41)$$

(here we assume that $c(\xi) = 0$ for $\xi \notin B_m$ (see [An7], §3.2)). We remark that for $X \in \mathcal{L}_0$ not belonging to the subalgebra $\varepsilon_0(\mathcal{L}) \subset \mathcal{L}_0$ we have, in general, that $f|X \notin \mathcal{M}_m^k(N, \psi)$. However we claim that $f|X \in \mathcal{M}_m^k(\Gamma'_X)$, where

$$\Gamma'_X = \cap_i g_i^{-1} \Gamma_1^m(N) g_i$$

(see (1.19)).

As an example we consider the Frobenius elements (see [An7], §2.1):

$$\begin{aligned} \Pi_- = \Pi_-^m(q) &= (\Gamma_0 \begin{pmatrix} q1_m & 0 \\ 0 & 1_m \end{pmatrix} \Gamma_0), \\ \Pi_+ = \Pi_+^m(q) &= (\Gamma_0 \begin{pmatrix} 1_m & 0 \\ 0 & q1_m \end{pmatrix} \Gamma_0). \end{aligned} \tag{1.42}$$

Then for $f(z) = \sum_{\xi \in B_m} c(\xi) e_m(\xi z) \in \mathcal{M}_m^k(N, \psi)$ we get from (1.41)

$$(c|\Pi_-)(\xi) = \psi(q)^m q^{mk - \langle m \rangle} c(q^{-1}\xi), \tag{1.43}$$

$$(c|\Pi_+)(\xi) = c(q\xi) \quad (\xi \in B_m \subset A_m) \tag{1.44}$$

Note that the operator $\Pi_+ = \Pi_+(q)$ is defined by the formula (1.44) also for $q|N$, when $\Pi_+(q)$ sends $\mathcal{M}_m^k(N, \psi)$ to itself. However we assume in the next proposition that $q \nmid N$.

1.9. Proposition. *If $f \in \mathcal{M}_m^k(N, \psi)$ then for a fixed prime q we have that*

$$f|\Pi_-(q), f|\Pi_+(q) \in \mathcal{M}_m^k(qN, \psi). \tag{1.45}$$

Proof. Note that

$$\Pi_-^m(q) = (\Gamma_0 \begin{pmatrix} q1_m & 0 \\ 0 & 1_m \end{pmatrix}),$$

therefore

$$f|\Pi_-(q) = f|_{k,\psi} \begin{pmatrix} q1_m & 0 \\ 0 & 1_m \end{pmatrix},$$

If we now take into account that

$$\Gamma_0^m(qN) \subset \begin{pmatrix} q1_m & 0 \\ 0 & 1_m \end{pmatrix}^{-1} \Gamma_0^m(N) \begin{pmatrix} q1_m & 0 \\ 0 & 1_m \end{pmatrix},$$

then we get that for all $\gamma \in \Gamma_0^m(qN)$ the following equality holds

$$f|\Pi_-\gamma = (f|\Pi_-)|_{k,\psi}\gamma = f|\Pi_-.$$

In order to prove the analogous statement about Π_+ we use the antiisomorphism (involition) $X \mapsto X^*$ of the algebra $\mathcal{L}_0$ (see [An7], p.75) which has the property $\Pi_-^* = \Pi_+$. We recall that by definition if

$$X = \sum_{i=1}^{t(X)} v_i(\Gamma_0 g_i) \in \mathcal{L}_0 \quad (v_i \in \mathbf{Q}, g_i \in \Delta_0),$$

then

$$X^* = \sum_{i=1}^{t(X)} v_i(\Gamma_0 q^{\nu_i} g_i^{-1}) \quad (q^{\nu_i} = \nu(q).$$

Next, let the equality $X\Gamma' = X$ be valid for a given $X \in \mathcal{L}_0$ where $\Gamma_0 \subset \Gamma'\Gamma$. Put $\Delta' = \Gamma'\Delta_0$, $\Delta_0 \subset \Delta' \subset \Delta$. Consider as in 1.3 the Hecke algebra $\mathcal{L}' = \mathcal{D}_{\mathbf{Q}}(\Gamma', \Delta')$. Then by definition of the homomorphism $\varepsilon' : \mathcal{L}' \to \mathcal{L}_0$ we have again that its image $\varepsilon_0'(\mathcal{L}')$ coincides with the subalgebra of all elements in $\mathcal{L}_0$ fixed by Γ'. Consequently, for some $Y \in \mathcal{L}'$ we have that $X = \varepsilon_0'(Y)$ and we get $X^* = \varepsilon_0'(Y^*)$ by the definition of the antiautomorphism $*$. In other words, the operator $X^* \in \mathcal{L}'$ turns out to be Γ'-invariant. In particulary we get $\Pi_+\Gamma_0^m(Nq) = \Pi_+$ by the already proven equality $\Pi_-\Gamma_0^m(Nq) = \Pi_-$, and by the fact that $f|\Pi_+ \in \mathcal{M}_m^k(qN, \psi)$.

Now let us consider again the Hecke polynomials, defined by (1.22) – (1.24). A. N. Andrianov has dicoverd factorization of the polynomials $Q(z), R(z) \in \mathcal{L}[z]$ over the non - commutative algebra $\mathcal{L}_0 = \mathcal{L}_{0,q}^m$ in terms of the Frobenius elements $\Pi_+(q), \Pi_-(q)$ (see 1.8). This factorization is essentially used in our non-Archimedean construction.

1.10. Proposition. *There are the following factorizations of the Hecke polynomials introduced above:*

$$Q(z) = (\sum_{i=1}^{2^m-1} V_i^+ z^i)(1 - \Pi_+ z) = (1 - \Pi_- z)(\sum_{i=1}^{2^m-1} V_i^- z^i), \tag{1.46}$$

where

$$V_i^+ = \sum_{j=0}^{i} (-1)^j T_j \Pi_+^{i-j},$$
$$V_i^- = \sum_{j=0}^{i} (-1)^j \Pi_-^{i-j} T_j \in \mathcal{L}_0 \tag{1.46a}$$

Proof (see in [An7], §2.2)

Now let

$$f(z) = \sum_{\xi \in B_m} c(\xi) e_m(\xi z) \in \mathcal{M}_m^k(N, \psi)$$

be an eigenfunction of the algebra $\mathcal{L} = \mathcal{L}_p^m(N)$ whose eigenvalue is a homomorphism $\lambda_f : \mathcal{L} \to \mathbf{C}, f|X = \lambda_f(X)f$ for all $X \in \mathcal{L}$ which is given by the $(m+1)$-tuple $\alpha_0 = \alpha_0(p), \alpha_1 = \alpha_1(p), \cdots, \alpha_m = \alpha_m(p)$ of the Satake p-parameters. Put

$$f_0 = \sum_{i=1}^{\tilde{m}-1} \alpha_0(p)^{-i} f|V_i^+(p) \quad (\tilde{m} = 2^m). \tag{1.47}$$

1.11. Proposition. a). *The function f_0 belongs to $\mathcal{M}_m^k(Np^{\tilde{m}-1}, \psi)$ with $\tilde{m} = 2^m$.*

b). *The following equality holds*

$$f_0|\Pi_+(p) = \alpha_0(p) f_0.$$

Remark. Using the action of the Weyl group in the construction of Proposition 1.11 we can obtain 2^m eigenfunctions $g \in \mathcal{M}^k_m(Np^{\tilde{m}-1}, \psi)$ of the local Hecke algebras $\mathcal{L} = \mathcal{L}^m_q(N)$ for $q \not| pN$ with the same eigenvalues as those of the original modular form $f \in \mathcal{M}^k_m(N, \psi)$ such that for $q = p$ we have that $g|\Pi_+(p) = \alpha g$ where α is any of the inverse roots

$$\alpha_0\alpha_{i_1}\cdots\alpha_{i_r} \quad (1 \le i_1 < \cdots \le i_r \le m)$$

of the local factor $Z^{(p)}(s, f)$ of the spinor zeta function.

As was pointed out to the author by S.Böcherer, these functions may vanish if we take for f a certain Eisenstein seies (see the remark before the Main theorem in §1 of Chapter 3). He mentioned also that a stronger statement about the level of the above form f_0 is valid. However we do not give here this interesting result because it is not directly needed for our proof.

Proof of the Proposition 1.11. The statement a) follows directly from (1.46a) and from the definition (1.47). In order to prove b) we note that there is the equality (with $\tilde{m} = 2^m$)):

$$\sum_{i=0}^{\tilde{m}=2^m} f|T_i(p)\cdot(-z)^i = \tilde{Q}(\alpha_0(p), \alpha_1(p)\cdots, \alpha_m(p); z)f;$$

in addition by definition of the polynomials (1.22) we have the identity:

$$\sum_{i=0}^{\tilde{m}=2^m} (-\alpha_0(p))^{-i} f|T_i(p) = 0. \tag{1.48}$$

We want to prove that $f_0|(\Pi_+(p) - \alpha_0(p)) = 0$. But we have from (1.46),(1.47) that

$$f_0|(\Pi_+(p) - \alpha_0(p)) = -\alpha_0(p)f|(1 - \alpha_0^{-1}(p)\Pi_+(p)) =$$

$$-\alpha_0(p)f|(\sum_{i=0}^{\tilde{m}=2^m} (-\alpha_0(p))^{-i} f|V_i^+(p))(1 - \alpha_0^{-1}(p)\Pi_+(p)) =$$

$$\sum_{i=0}^{\tilde{m}=2^m} f|T_i(p)(-\alpha_0(p))^{-i} = 0,$$

and the statement b) follows.

1.12. An analogous statement is valid in the case of a finite set S of prime numbers q coprime to N. For this purpose it is convenient to consider the global Hecke algebras

$$\mathcal{L}(N) = \otimes_{q \not| N}\mathcal{L}^m_q(N), \quad \mathcal{L}_0(N) = \otimes_{q \not| N}\mathcal{L}^m_{0,q}(N).$$

Then the definition of the operators $\Pi_+(M), \Pi_-(M), V_i^+(q), V_i^-(q) \in \mathcal{L}^m_{0,q}(N)$ and of the numbers $\alpha_0(q)$ $(q \not| N)$ can be extended by multiplicativity to all positive integers M coprime with N; more precisely, operators $V^+(M), V^-(M)$ are defined by the identities

$$\sum_{M|M_0^{\tilde{m}-1}} M^{-s}V^+(M) = \prod_{q \not| N} [\sum_{i=1}^{\tilde{m}-1} q^{-s}V_i^+(q)], \tag{1.49}$$

$$\sum_{M|M_0^{\tilde{m}-1}} M^{-s}V^-(M) = \prod_{q\not| N} [\sum_{i=1}^{\tilde{m}-1} q^{-s}V_i^-(q)], \tag{1.49a}$$

where the notation

$$V^{\pm}(q^i) = V_i^{\pm}(q), \quad M_0 = \prod_{q\in S} q.$$

is used. We then put

$$f_0 = f_{0,S} = \sum_{M|M_0^{\tilde{m}-1}} \alpha_0(M)^{-1} f|V^+(M). \tag{1.50}$$

1.13. Proposition. *a) Let* $f \in \mathcal{M}_k^m(N,\psi)$, *then*

$$f_0 = f_{0,S} \in \mathcal{M}_k^m(NM^{\tilde{m}-1},\psi).$$

b) For all positive integers M *with support* $S(M)$ *in the set* S *we have that*

$$f_0|\Pi_+(M) = \alpha_0(M)f_0. \tag{1.51}$$

c) Let

$$f_{0,S}(z) = \sum_{\xi\in B_m} a_0(\xi)e_m(\xi z) \in \mathcal{M}_m^k(NM^{\tilde{m}-1},\psi) \quad (\xi \in B_m)$$

be Fourier expansion of the function $f_{0,S}(z)$ *then there is the following multiplicativity property of its Fourier coefficients: for all* $M \in \mathbf{N}$ *with* $S(M) \subset S$ *we have that*

$$a_0(M\xi) = \alpha_0(M)a_0(\xi) \quad (\xi \in A_m, \xi \geq 0). \tag{1.52}$$

The proof of the proposition is carried out in a very similar way as that of the previous one if we take into account formulas (1.44) for the action of $\Pi_+(M)$ on Fourier expansions.

§2. Theta series, Siegel – Eisenstein series and the Rankin zeta function

2.1. Theta series (see [An3], [An-M1], [An-M2], [St2]). Let $F \in 2C_m$ be an even symmetric positive definite matrix, and q_0 its level (i.e. the smallest positive integer such that $q_0F^{-1} \in 2C_m$, and χ a Dirichlet character modulo Q (not necesarily primitive). Put $\nu = 1$ or 0 and define the theta function

$$\theta(\chi) = \theta_F^{(\nu)}(z;\chi) = \sum_{\xi\in \mathrm{M}_m(\mathbf{Z})} \chi(\det\xi)\det\xi^{\nu} e_m(zF[\xi]/2). \tag{2.1}$$

2.2. Proposition. *(a) If*

$$\gamma = \begin{pmatrix} a & b \\ c & d \end{pmatrix} \in \Gamma_0^m(q_0Q^2),$$

then the following transformation formula holds

$$\theta_F^{(\nu)}(\gamma(z);\chi) = \chi(\det d)\chi_{Q^2F}^{(m)}(\gamma)\det(cz+d)^{(m/2)+\nu}\theta_F^{(\nu)}(z;\chi), \tag{2.2}$$

where $\chi_F^{(m)}(\gamma)$ is a root of unity of the eighth degree, and if m is even, then

$$\chi_F^{(m)}(\gamma) = \left(\frac{(-1)^{m/2}\det(F)}{\det d}\right). \tag{2.3}$$

(b) Let $J(M)$ denote for $M>0$ the matrix $\begin{pmatrix} 0_m & -1_m \\ M1_m & 0_m \end{pmatrix}$. If χ is primitive modulo Q then the action of the involution $J(Q^2q_0)$ on (2.1) is given by

$$\begin{aligned}\theta_F^{(\nu)}(J(Q^2q_0)z;\chi) &= \\ &\chi(-1)^m Q^{m\nu} G_Q(1_m,\chi)\det(F)^{(m/2)+\nu}[\det(-iz)]^{(m/2)+\nu}\theta_{\hat{F}}^{(\nu)}(z;\overline{\chi}),\end{aligned} \tag{2.4}$$

with $\hat{F} = q_0F^{-1}$ and

$$G_Q(\xi,\chi) = \sum_{h\in \mathrm{M}_m(\mathbf{Z}) \bmod Q} \chi(\det h)\, e_m({}^t\xi h/Q)$$

is the Gauss sum of degree m of the character χ.

2.3. *The proof* of (a) is given in [An-M2] for primitive characters χ, and in [St2] in the general case by use of the generalized theta series

$$\theta_F^{(m)}(z;X,Y) = \sum_{\xi\in \mathrm{M}_m(\mathbf{Z})} e_m((zF[\xi - Y] + 2^\xi X - {}^tXY)/2), \tag{2.6}$$

where $z \in \mathfrak{H}_m$, $X, Y \in \mathrm{M}_m(\mathbf{C})$. The series (2.6) satisfy the following properties: for all $\gamma = \begin{pmatrix} a & b \\ c & d \end{pmatrix} \in \Gamma_0^m(q_0)$ we have that

$$\begin{aligned}\theta_F^{(m)}(\gamma(z); X^a + FY^tb, F^{-1}X^tc + Y^td) &= \\ &\chi_F^{(m)}(\gamma)\det(cz+d)^{(m/2)}\theta_F^{(m)}(z;X,Y),\end{aligned} \tag{2.7}$$

$$\theta_F^{(m)}(z;X,Y) = (\det F)^{-(m/2)}[\det(-iz)]^{-(m/2)}\theta_{F^{-1}}^{(m)}(-z^{-1};Y,-X) \tag{2.8}$$

(see [An-M2], theorem 1 and lemma 2). For $\nu = 0$ the statements (a) and (b) follow immediately from (2.7), (2.8). For example, in order to prove (b) it suffices to put in (2.8) F equal to Q^2F, $X=0$, Y equal to $-Q^{-1}Y$ then multiply the result by $\chi(\det(Y))$ and to carry out summation over $Y \in \mathrm{M}_m(\mathbf{Z}) \bmod Q$. From the left hand side of (2.8) we get

$$\begin{aligned}\sum_{Y\in \mathrm{M}_m(\mathbf{Z}) \bmod Q} &\chi(\det(Y))\theta_{Q^2F}^{(m)}(z;0,-Q^{-1}Y) = \\ &\sum_{Y\in \mathrm{M}_m(\mathbf{Z}) \bmod Q} \chi(\det(Y)) \sum_{\xi\in \mathrm{M}_m(\mathbf{Z})} e_m(zF[Q\xi + Y]/2),\end{aligned} \tag{2.9}$$

and from the right hand side

$$(\det Q^2F)^{-(m/2)}[\det(-iz)]^{-(m/2)}\theta^{(m)}_{Q^{-2}F^{-1}}(-z^{-1};-Q^{-1}Y,0) =$$
$$(\det Q^2F)^{-(m/2)}[\det(-iz)]^{-(m/2)}\times$$
$$\times \sum_{Y\in \mathrm{M}_m(\mathbf{Z}) \bmod Q} \chi(\det(Y)) \sum_{\xi\in \mathrm{M}_m(\mathbf{Z})} e_m((-(zQ^2q_0)^{-1}(q_0F^{-1})[\xi] - 2Q^{-1\,t}\xi Y)/2). \tag{2.10}$$

Now we note that for any Dirichlet character χ the definition (2.1) can be rewritten as follows:

$$\theta(\chi) = \theta^{(\nu)}_F(z;\chi) = \sum_{Y\in \mathrm{M}_m(\mathbf{Z}) \bmod Q} \chi(\det(Y)) \sum_{\xi\equiv Y \bmod Q} (\det\xi)^\nu e_m(zF[\xi]/2). \tag{2.11}$$

If we take into account the primitivity of χ then we get (see [An5], (5.12))

$$\theta^{(\nu)}_F(z;\chi) = G_Q(1_m,\overline{\chi})^{-1}\times \sum_{Y\in \mathrm{M}_m(\mathbf{Z}) \bmod Q} \overline{\chi}(\det(Y)) \sum_{\xi\equiv Y \bmod Q} (\det\xi)^\nu e_m(zF[\xi] + 2Q^{-1\,t}\xi Y)/2), \tag{2.12}$$

Let us use the standard relation

$$G_Q(1_m,\chi)G_Q(1_m,\overline{\chi}) = \chi(-1)Q^{m^2}. \tag{2.13}$$

for Gauss sums. We see that (2.9) becomes

$$\theta^{(0)}_F(z;\chi), \tag{2.14}$$

and (2.10) transforms to the following

$$\begin{aligned}&(\det Q^2F)^{-(m/2)}[\det(-iz)]^{-(m/2)}G_Q(1_m,\chi)\theta^{(0)}_{q_0F^{-1}}(J(q_0Q^2)(z);\overline{\chi}) =\\ &\chi(-1)(\det F)^{-(m/2)}[\det(-iz)]^{-(m/2)}G_Q(1_m,\overline{\chi})\theta^{(0)}_{q_0F^{-1}}(J(q_0Q^2)(z);\overline{\chi}),\end{aligned} \tag{2.15}$$

Let us set in (2.12) F equal to q_0F^{-1}, z equal to $-(Q^2q_0z)^{-1}$, replace χ by $\overline{\chi}$ and ξ by $-\xi$; comparison of (2.14) and (2.15) provides us then with an identity, which is equivalent to (2.4) with $\nu = 0$.

In order to prove the statements (a) and (b) in the case $\nu = 1$ we take a matrix $F_1 = {}^tF_1 \in \mathrm{M}_m(\mathbf{R})$ such that $f_1 > 0, F_1^2 = F$, and let $\eta = (\eta_{ij}) \in \mathrm{M}_m(\mathbf{C})$ be a matrix variable. If we set in (2.7) $X = 0, Y = F_1^{-1}\eta$ and apply the differential operator $L_\eta = \det(\partial/\partial\eta_{ij})$ to both parts of the resulting equality, then after some simplifications we get the equality (2.2) with $\nu = 1$. The differential identity

$$\begin{aligned}&L_\eta(e_m((P\,{}^t\eta\eta + 2\,{}^tT\eta + R)/2)) =\\ &\det(2\pi i(\eta P + T))e_m((P\,{}^t\eta\eta + 2\,{}^tT\eta + R)/2)),\end{aligned} \tag{2.16}$$

is particulary useful in this calculation, where $P, T, R \in \mathrm{M}_m(\mathbf{C})$, ${}^tP = P$ (see [An7], lemma 5.1).

We give a more detailed proof of (b). Let us set in (2.8) $X = 0$ and $Y = F_1^{-1}\eta$ and apply the operator L_η. Then the expression in the exponent of the summand corresponding to an integral matrix ξ in the left hand side of (2.8) becomes

$$\pi i \cdot \mathrm{tr}(zF[\xi - F_1^{-1}\eta]) = \pi i \cdot \mathrm{tr}(z\,{}^t\eta\eta - 2\,{}^t(F_1\xi z)\eta + zF[\xi]).$$

This is easily deduced from the definition (2.6) if we notice that

$$\mathrm{tr}(AB) = \mathrm{tr}(BA) = \mathrm{tr}({}^tA\,{}^tB).$$

Let us apply to this term the operator L_η. According to (2.16) this term will then be multiplied by

$$\det(2\pi i(\eta z - F_1\xi z)) = (2\pi i)^m \det F^{1/2} \det((F_1^{-1}\eta - \xi)z).$$

Similarly, after application of L_η to each summand corresponding to a matrix ξ in

$$(\det F)^{-(m/2)}[\det(-iz)]^{-(m/2)}\theta^{(m)}_{F^{-1}}(-z^{-1}; F_1^{-1}\eta, 0),$$

this summand will get the factor $(\det F)^{-1/2}(2\pi i)^m \det(\xi)$. If we remember that $Y = F_1^{-1}\eta$, then we see that the result of the action of L_η on the expression (2.8) will be given by the following:

$$\begin{aligned} \det(z) \sum_{\xi \in \mathrm{M}_m(\mathbf{Z})} & \det(Y - \xi) e_m(zF[\xi - Y]/2) = \\ & (\det F)^{-(m/2)}[\det(-iz)]^{-(m/2)} \times \\ & \times \sum_{\xi \in \mathrm{M}_m(\mathbf{Z})} \det(\xi) e_m((-z^{-1}F^{-1}[\xi] + 2\,{}^t\xi Y)/2). \end{aligned} \tag{2.17}$$

Now again we set in (2.17) F equal to Q^2F, Y equal to $-Q^{-1}Y$, ξ equal to $-\xi$, multiply by $\chi(\det(Y))$ and carry out summation over Y mod Q keeping in mind the relation (2.13); as a result we get the equality (2.4) with $\nu = 1$. An explicit calculation of the multiplier $\chi_F^{(m)}(\gamma)$ for even m is given, for example, in [An-K].

2.4. Siegel – Eisenstein series. We start by recalling the definition of these series. We call matrices $c, d \in \mathrm{M}(\mathbf{Z})$ coprime if

$$\{G \in \mathrm{M}(\mathbf{Q}) | Gc,\ Gd \in \mathrm{M}(\mathbf{Z})\} = \mathrm{M}(\mathbf{Z})$$

A couple (c, d) is called a *symmetric couple* if $c\,{}^td = d^tc$. Two symmetric couples of coprime matrices are called *equivalent* iff for some unimodular matrix $U \in \mathrm{GL}_m(\mathbf{Z})$ we have $(c_1, d_1) = (Uc_2, Ud_2)$.

Let $\Delta = \Delta_m$ denote the set of equivalence classes of symmetric couples of coprime matrices. Then the set can be identified with the set of right coset classes $\Gamma_0^m\backslash\Gamma^m$ of the group $\Gamma^m = \mathrm{Sp}_m(\mathbf{Z})$ with respect to its parabolic subgroup

$$\Gamma_0^m = \left\{\gamma = \begin{pmatrix} a & b \\ 0 & d \end{pmatrix} | \gamma \in \Gamma^m\right\}$$

via the map

$$\Gamma_0^m \backslash \Gamma^m \ni \Gamma_0^m \begin{pmatrix} a & b \\ c & d \end{pmatrix} \mapsto \text{class of } (c,d) \in \Delta_m. \tag{2.18}$$

We see also that via this map the set

$$\{(c,d) \in \Delta_m | c \equiv 0(\text{mod} N)\}$$

is identified with a coset for $\Gamma_0^m \backslash \Gamma_0^m(N)$.

Now let k , N be positive integers, s a complex number and χ a Dirichlet character modulo N such that $\chi(-1) = (-1)^k$. For $z \in \mathfrak{H}_m$ the Siegel – Eisenstein series is defined by

$$E(z,s;k,\chi,N) = E(z,s) = \det(y)^s \sum \chi(\det(d)) \det(cz+d)^{-k-|2s|}, \tag{2.19}$$

where the summation is taken over all $(c,d) \in \Delta$ with the condition $c \equiv 0(\text{mod} N)$ and we adopt the convenient notation by Deligne and Ribet [De-R]:

$$z^{-k-|2s|} \stackrel{\text{def}}{=} z^{-k}|z|^{-2s} \text{ for } z \in \mathbf{C}^*$$

The series (2.19) is absolutely convergent for $k + 2\text{Re}(s) > m + 1$ and it admits a meromorphic analytic continuation over the whole complex s-plane. Put $j(\alpha, z) = \det(cz + d)$ for $\alpha = \begin{pmatrix} a & b \\ c & d \end{pmatrix}$ and $z \in \mathfrak{H}_m$, then it follows from the description of Δ_m given above that

$$E(z,s;k,\chi,N) = \det(y)^s \sum_{\alpha \in P \cap \Gamma \backslash \Gamma} \chi(\det(d_\alpha) j(\alpha,z)^{-k-|2s|}, \tag{2.20}$$

where $\Gamma = \Gamma_0^m(N), \alpha = \begin{pmatrix} a_\alpha & b_\alpha \\ c_\alpha & d_\alpha \end{pmatrix}$, and P denotes the subgroup of $P \subset G_{\infty+}$, consisting of elements α with the condition $c_\alpha = 0$.

2.5. The Rankin zeta function. Let f and g be two holomorphic modular forms of weights k and l on the congruence subgroup $\gamma_0^m(N) \subset \Gamma^m$, and with Dirichlet characters ψ and ω. More precisely we suppose that

$$f(z) = \sum_{\xi \in B_m} a(\xi) e_m(\xi z) \in \mathcal{S}_m^k(N, \psi), \tag{2.21}$$

$$g(z) = \sum_{\xi \in C_m} b(\xi) e_m(\xi z) \in \mathcal{M}_m^l(N, \psi), \tag{2.22}$$

with B_m being the set of half integral non negative matrices of size $m \times m$, and $C_m \subset B_m$ the subset of all positive definite matrices, see §1. Define an equivalence relation on B_m by $\xi_1 \sim \xi_2$ iff $\xi_1 = {}^t u \xi_2 u$ for some matrix $u \in \text{SL}_m(\mathbf{Z})$. Then the *Rankin zeta function* (convolution of two Siegel modular forms) is defined as the series [St2]:

$$L(s,f,g) = \sum_{\xi \in \tilde{C}_m} \varepsilon(\xi)^{-1} a(\xi) b(\xi) \det(\xi)^{-s} \quad (s \in \mathbf{C}), \tag{2.23}$$

where $\varepsilon(\xi) = |\mathrm{Aut}(\xi)$is the order of the unit group

$$\mathrm{Aut}(\xi) = \{\eta \in \mathrm{SL}_m(\mathbf{Z}) |\ {}^t\eta\xi\eta = \xi[\eta] = \xi\}.$$

This series is well defined due to basic properties of Fourier coefficients of Siegel modular forms; $\tilde{C}_m = C_m \mathrm{mod} \sim$ denotes the orbit space with respect to $\sim$ and $\mathrm{Re}(s)$ is supposed to be large enough: $\mathrm{Re}(s) \gg 0$. To be more precise, we note that the series

$$\sum_{\xi \in \tilde{C}_m} \det(\xi)^{-s} \tag{2.24}$$

is absolutely convergent for $\mathrm{Re}(s) > m$. It is easily seen from the fact that cardinality of the set

$$\mathrm{SL}_m(\mathbf{Z}) \backslash \{\gamma \in \mathrm{M}_m(\mathbf{Z}) | \det\gamma = a > 0\}$$

is estimated by the finite sum $\sum a_2 a_3^2 \cdots a_m^{m-1}$, hence the Dirichlet series in (2.24) admits the upper bound

$$\prod_{i=0}^{m-1} \zeta(\mathrm{Re}(s) - i),$$

with $\zeta(s)$ being the Riemann zeta function (see [An7], p.133). Keeping in mind estimates (1.15a) and (1.16a) for Fourier coefficients of Siegel modular forms we get

$$a(\xi) = \mathcal{O}(\det(\xi)^{k/2}), \quad b(\xi) = \mathcal{O}(\det(\xi)^l),$$

so that the Dirichlet series (2.23) is absolutely convergent for $\mathrm{Re}(s) > m + k/2 + l + \varepsilon$, $(\varepsilon > 0)$.

2.6. Proposition. *(Integral representation for the Rankin zeta function, see [St2], proposition 6).*

For s with $\mathrm{Re}(s) \gg 0$ there is the following integral representation.

$$(4\pi)^{-ms}\Gamma_m(s)L(s, f, g) = \langle f^\rho, gE(z, s - k + (m+1)/2; k - l, \psi\omega, N)\rangle_N \tag{2.25}$$

where the inner product is defined by (1.13),

$$f^\rho(z) = \sum_{\xi \in B_m} \overline{a(\xi)} e_m(\xi z) \in \mathcal{S}_m^k(N, \psi),$$

and $\Gamma_m(s)$ denotes the Γ-function of degree m, i.e.

$$\Gamma_m(s) = \pi^{m(m-1)/4} \prod_{j=0}^{m-1} \Gamma(s - (j/2))$$

(see also (3.8)).

2.7. The standard zeta function $\mathcal{D}(s, f, \chi)$ as the Rankin convolution. Now let

$$f(z) = \sum_{\xi \in C_m} \overline{a(\xi)} e_m(\xi z) \in \mathcal{S}_m^k(N, \psi)$$

be a cusp form of even degree m which is an eigenfunction of the Hecke algebra $\mathcal{L}_q^m(N)$ for q not dividing N, and let

$$(\alpha_0, \alpha_1, \cdots, \alpha_m) = (\alpha_0(q), \alpha_1(q), \cdots, \alpha_m^q)) \in [(A^\times)^{m+1}]^{W_m}$$

be the corresponding $(m+1)$-tuple of the Satake q-parameters of f (see (1.18)). Fix a matrix $\xi_0 \in C_m$ such that $a(\xi_0) \neq 0$ and consider the primitive quadratic Dirichlet character χ_{ξ_0} definied for positive integers d with $(d, 2\det(2\xi_0)) = 1$ by the formula

$$\chi_{\xi_0}(d) = \left(\frac{(-1)^{m/2}\det(2\xi_0)}{d}\right),$$

and when $\det(2\xi_0)$ is odd, then by definition $\chi_{\xi_0}(2) = 1$ or -1 according as to which of the two following quadratic forms

$$x_1x_2 + x_3x_4 + \ldots x_{m-1}x_m$$

or

$$x_1x_2 + \ldots + x_{m-3}x_{m-2} + x_{m-1}^2 + x_{m-1}x_m + x_m^2$$

our quadratic form ξ_0 is equivalent over the field $\mathbf{F}_2$ of two elements. Assume also that $\chi \bmod M$ is chosen so that $(-1)^\nu = \chi(-1)$. The main result of the A.N.Andrianov's work [An6] can be stated as a certain identity expressing the standard zeta function $\mathcal{D}(s, f, \chi)$ as a Rankin zeta function, namely the convolution of the given form f and a theta function with the Dirichlet character $\chi \bmod M$. The precise statement of the result is given in the following proposition.

2.8. Proposition. *Under the notation and assumptions as above for the sufficiently large values of* $\mathrm{Re}(s)$ *the following identity holds*

$$\begin{aligned} &a(\xi_0)R(s, f, \chi) = \\ &\quad 2^{-1}\det \xi_0^{(s+k-1+\nu)/2} L((s+k-1+\nu)/2, f, \theta_{2\xi_0}^{(\nu)}(z;\chi)), \end{aligned} \tag{2.26}$$

where the function $R(s, f, \chi)$ *is defined by the following equality*

$$\mathcal{D}(s, f, \chi) = L(s + (m/2), \psi\chi_{\xi_0}\chi) \prod_{i=0}^{m-1} L(2s + 2i, \psi^2\chi^2) R(s, f, \chi), \tag{2.27}$$

and it is assumed that the modulus M of the character χ is divisible by all prime divisors of the number $N\det 2\xi_0$, where $\theta_{2\xi_0}^{(\nu)}(z;\chi)$ is the theta function as in 2.1.

More explicitly, the right hand side of (2.26) can be represented as the series

$$\begin{aligned} &(\det\xi_0)^{(s+k-1+\nu)/2} \sum_\xi \chi(\det\xi)\det\xi^\nu a(\xi_0[\xi])(\det\xi_0[\xi])^{-(s+k-1+\nu)/2} = \\ &\sum_\xi \chi(\det\xi) a({}^t\xi\xi_0\xi)(\det\xi)^{-(s+k-1+\nu)/2}, \end{aligned} \tag{2.28}$$

where the summation is taken over the set of equivalence classes $\xi \in \mathrm{SL}_m(\mathbf{Z})\backslash \mathrm{M}_m^+(\mathbf{Z})$ of the form $\mathrm{SL}_m(\mathbf{Z})\xi$.

Now we set q_0 to be equal to the level of the quadratic form with the matrix $2\xi_0$ (see 2.1). In order to get the integral representation of the standard zeta function we apply to it the result of 2.6. For this purpose we put in the notation of 2.6 $l = (m/2)+\nu$, $\omega = \chi\chi_{\xi_0}$ with χ being a Dirichlet character modulo M, and take the number Nq_0M^2 as N. We note also that both parts of (2.27) and (2.26) converge absolutely for $\mathrm{Re}(s) > m$ (see also [An7], p. 133).

2.9. Proposition *(integral representation of the standard zeta function). With the notation and assumption as above for* Re(s) *the following equality holds:*

$$\begin{aligned} &2a(\xi_0)(\det\xi_0)^{-(s+k-1+\nu)/2}(4\pi)^{-ms} \times \Gamma_m((s+k-1+\nu)/2)\, R(s,f,\chi) = \\ &\langle f^\rho, \theta^{(\nu)}_{2\xi_0}(z;\chi)E(z,(s-k+m+\nu)/2)\rangle_{NM^2q_0}, \end{aligned} \tag{2.29}$$

where the Siegel – Eisenstein series in the right hand side

$$E(z,s) = E(z,s;k-\nu-(m/2),\chi\chi_{\xi_0}\psi, NM^2q_0)$$

is defined by (2.29) .

§3. Formulas for Fourier coefficients of the Siegel – Eisenstein series

3.1. Rationality properties of Fourier coefficients. For the full symplectic modular group $\Gamma = \mathrm{Sp}_m(\mathbf{Z})$ Siegel has defined the series [Sie2], [Sie3] (see §2, 2.4)

$$E(z) = E_k^{(m)}(z) = \sum_{\gamma\in P\cap\Gamma^m\backslash\Gamma^m} j(\gamma,z)^{-k}, \tag{3.1}$$

where $z \in \mathfrak{H}_m$ is the point of the Siegel upper half plane of degree m, $j(\gamma,z) = \det(cz+d)$ for $\gamma = \begin{pmatrix} a & b \\ c & d \end{pmatrix} \in \Gamma^m$, P denotes the subgroup of Γ^m consisting of block matrices of the form $\begin{pmatrix} * & * \\ 0 & * \end{pmatrix}$. In the original definition by Siegel the number k is even and $k > m+1$ so that the series (3.1) is absolutely convergent and is referred to as Siegel – Eisenstein series. The rationality property of its Fourier coefficients were established by Siegel himself (although it was certainly known earlier in the case $m = 1$:

$$\begin{aligned} &E_k^{(1)}(z) = 1 - \frac{2k}{B_k}\sum_{n=1}^{\infty}\sigma_{k-1}(n)e(nz) = \\ &1 + \sum_{n=1}^{\infty}\left(\frac{2\sigma_{k-1}(n)}{\zeta(1-k)}\right)e(nz), \quad \sigma_{k-1}(n) = \sum_{d\mid n} d^{k-1}, \end{aligned} \tag{3.2}$$

where B_k are Bernoulli numbers, $\zeta(s)$ being the Riemann zeta function.) After the original Siegel's work his calculation was generalized in various directions: to the case of congruence subgroups of $\Gamma_0^{(m)}(N) \subset \Gamma^m$ [St2] to non-convergent series defined by analytic continuation over an additional parameter (Hecke's method) [He], [Fe], to other classes of algebraic groups and symmetric domains [Ba], [Har2], [Har4], [Fe], [Shi7], [Shi10]. It was discovered that the rationality property remains valid even for more complicated series, which themselves are defined by some inductive process of inducing from other cusp forms of lower degree (the *Klingen – Eisenstein series*). More precisely, let $f \in \mathcal{S}_k^r$ be a cusp form of degree r (with respect to the group Γ^r). If $k > m + r + 1$ and $m \geq r$ then the Klingen – Eisenstein series is defined as the following absolutely convergent series

$$E_k^{m,r}(z, f) = \sum_{\gamma \in \Delta_{m,r} \backslash \Gamma^m} f((\gamma z)^{(r)}) j(\gamma, z)^{-k}, \tag{3.3}$$

with $z \in \mathfrak{H}_m$, $z^{(r)}$ being the upper left corner of z of size $r \times r$, and $\Delta_{m,r}$ denotes the set of elements in Γ^m having the form $\begin{pmatrix} * & * \\ 0_{m-r,m+r} & * \end{pmatrix}$ [Kl3]. This series turns out to be a modular form of degree m on the group Γ^m. M. Harris proved in [Har2] – [Har4] the validity of the Garrett's conjecture: all the Fourier coefficients the modular form $E_k^{m,r}(z, f)$ belong to the field $\mathbf{Q}(f)$ generated by the Fourier coefficients of f. Explicit formulas for Fourier coefficients of the series $E_k^{m,r}(z, f)$ were given by Böcherer [Bö2] in general for $k > m + r + 1$ and by Kurokawa and Mizumoto [Kur-Miz], [Miz1], [Miz2] who treated the case $m = 2, r = 1$. It turned out that the most significant term in these formulas involves the special values of the standard zeta function of f twisted with a certain quadratic Dirichlet character attached to the matrix index ξ of a Fourier coefficient; as it was noticed above, these functions reduce to the (twisted) symmetric squares of the form f if $m = 2$. The formulas of BÖcherer and Mizumota may be regarded as a vast generalization of the classical formulas (3.2), if we assume that cusp forms of degree 0 are constants and their zeta functions reduce to the Riemann zeta function and to Dirichlet L-series.

However, in what follows we are interested only in Siegel – Eisenstein series which were defined in §2 by:

$$E(z, s; k, \chi, N) = E(z, s) = \det(y)^s \sum_{\alpha \in P \cap \Gamma \backslash \Gamma} \chi(\det(d_\alpha) j(\alpha, z)^{-k-|2s|}, \tag{3.4}$$

for $k + 2\mathrm{Re}(s) > m + 1$ (we use again the notation $z^{-k-|2s|} = z^{-k}|z|^{-2s}$ for $z, s \in \mathbf{C}, k \in \mathbf{Z}$) and by analytic continuation over s for other values of $s \in \mathbf{C}$ (see [Kl], [Shi10] and 3.3 below). It is assumed in the identity (3.4) that $N > 1, \chi$ is a Dirichlet character mod N (not necessarily primitive, e.g. trivial modulo $N > 1$), and

$$\alpha = \begin{pmatrix} a_\alpha & b_\alpha \\ c_\alpha & d_\alpha \end{pmatrix} \in \Gamma = \Gamma_0^m(N) \subset \Gamma^m.$$

In the following study of certain arithmetic properties of Fourier coefficients we use an explicit calculation of the Fourier expansion of the series

$$E^*(z, s) = E(-z^{-1}, s)(\det z)^{-k}, \tag{3.5}$$

obtained from (3.4) by applying the involution

$$J_m = \begin{pmatrix} 0 & -1_m \\ 1_m & 0 \end{pmatrix}.$$

However for $k > m+1$ and $N = 1$ both series coincide and reduce to the series originally studied by Siegel:

$$E(z) = E_k^m(z) = E(z,0) = E^*(z,0).$$

The detailed study of these series was made by Shimura [Shi10] and P.Feit ([Fe], §10) in a more general situation, in particular, for the case of Eisenstein series attached to the group Sp_m over a totally real field. For convenience we reproduce only a specialization of these results to the case of $F = \mathbf{Q}$.

The remarkable summation method by means of the analytic continuation of the function $E(z,f)$ over s to the point $s = 0$ was first discovered by Hecke [He]. Quite recently this method was essentially extended by R.Weissauer [We2] to the case of Klingen – Eisenstein series (3.3) and then it was used for a construction of Siegel modular forms of "small" weight ($k = m, m+1$), which play a significant role in studying geometric invariants of Siegel modular varieties [We1],[We3].

3.2. Preparation: the confluent hypergeometric function. For a detailed description of the Fourier expansion of the series (3.5) we need some additional notation.Let

$$V = V_m = \{h \in \mathrm{M}_m(\mathbf{R}) | {}^t h = h\} \tag{3.6}$$

be the set of all real symmetric matrices of size $m \times m$, and

$$Y = V^+ \{h \in V | h > 0\} \tag{3.7}$$

the subset of its positive definite elements. For each matrix $T \in \mathrm{M}_m(\mathbf{R})$ let $\delta_+(T)$ denote the product of all positive eigenvalues of T, $\delta_-(T) = \delta_+(-T)$ and $\delta_+(T) = 1$ if T does not have positive eigenvalues.

For $h \in V$ let $p = p(h)$ denote the number of positive eigenvalues of h counted with their multiplicities, and $q = q(h)$ the number of negative eigenvalues. Then $r = r(h) = p + q$ is the rang of h.

Let also

$$\Gamma_m(s) = \pi^{m(m-1)/4} \prod_{j=0}^{m-1} \Gamma(s - (j/2)), \tag{3.8}$$

be the Γ-function of degree m. This function generalizes the usual Γ-function in view of the following integral representation

$$\Gamma_m(s) = \int_Y (\det y)^s e^{-\mathrm{tr}(y)} d^\times y, \tag{3.9}$$

which is valid for $s \in \mathbf{C}$ with $\mathrm{Re}(s) > (m-1)/2$, and

$$dy = \prod_{i \le j} dy_{ij}, \quad \det(y)^{-(m+1)/2} dy.$$

Recall that $d^\times y$ is a measure on Y which is invariant with respect to the action of $a \in \mathrm{GL}_m(\mathbf{R})$ given by $d^\times({}^t a y a) = d^\times y$. For complex numbers α and β we define the numbers

$$\kappa = (m+1)/2,$$

$$\tau = \tau(h,\alpha,\beta) = \\ = (2p-m)\alpha + (2q-m)\beta + m + (m-r)\kappa + pq/2, \tag{3.10}$$

$$\sigma = \sigma(h,\alpha,\beta) = \\ p\alpha + q\beta + m - r + \{(m-r)(m-r-1)\}/2. \tag{3.11}$$

In [Shi8] Shimura constructed the function

$$\omega(y,h;\alpha,\beta), \tag{3.12}$$

which is defined for all $(y,h;\alpha,\beta) \in Y \times V \times \mathbf{C}^2$ and is a holomorphic function of $(\alpha,\beta) \in \mathbf{C}^2$. It can be used for computing the Fourier expansion of the series

$$S(z,L;\alpha,\beta) = \sum_{a \in L} \det(z+a)^{-\alpha} \det(\overline{z}+a)^{-\beta} \quad (z \in \mathfrak{H}_m) \tag{3.13}$$

which is obtained by summation over a lattice $L \subset V$ and is absolutely convergent for $\mathrm{Re}(\alpha+\beta) > m$. Let

$$L' = \{h \in V | \mathrm{tr}(hL) \in \mathbf{Z}\}$$

be the lattice dual to L with respect to the pairing given by $(u,v) \mapsto e_m(uv)$. In particular there is the equality

$$\mu(V/L) S(z;\alpha,\beta) = \sum_{h \in L'} \xi(y,h;\alpha,\beta) e_m(hx) \tag{3.14}$$

in which

$$\mu(V/L) = \int_{V/L} dy$$

denotes the volume of the fundamental domain V/L,

$$\xi(y,h;\alpha,\beta) = \\ i^{m\beta - m\alpha} 2^{\tau} \pi^{\sigma} \Gamma_{m-r}(\alpha+\beta-\kappa) \Gamma_{m-q}(\alpha)^{-1} \Gamma_{m-p}(\beta)^{-1} \times \\ \times (\det y)^{\kappa-\alpha-\beta} \delta_{+}(hy)^{\alpha-\kappa+q/4} \delta_{-}(hy)^{\beta-\kappa+q/4} \omega(2\pi y, h;\alpha,\beta), \tag{3.15}$$

and is additionaly assumed that $\mathrm{Re}(\alpha) > m/2$, $\mathrm{Re}(\beta) > m/2$ (for the regularity of Γ-functions in (3.15) (see [Shi8], (4.34.K)) and we adopt the standard choice of branch for the exponentiation, namely,

$$v^{\alpha} = e^{\alpha \log(v)}, \quad -\pi \le \mathrm{Im}(\log v) < \pi.$$

The function $\xi(y,h;\alpha,\beta)$ admits the following integral representation: for $g \in Y, h \in V, (\alpha,\beta) \in \mathbf{C}^2$

$$\xi(y,h;\alpha,\beta) = \\ \int_V e_m(-hx) \det(x+ig)^{-\alpha} \det(x-ig)^{-\beta} dx, \tag{3.16}$$

with the integral being absolutely convergent for $\mathrm{Re}(\alpha+\beta) > 2\kappa+1$ (see [Shi8], (1.25)).

Applying the equality (3.14) to the lattice $L = S = V \cap \mathrm{M}_m(\mathbf{Z})$ when $L' = A = A_m$ is the lattice of all symmetric half integral matrices and also to the lattice $L = NS, L' = N^{-1}A_m$ with $\alpha = k, \beta = 0, k > m$ and $C_m = A_M \cap Y$ we get the classical equality:

$$\sum_{a \in S} \det(z+a)^{-k} = (2\pi i)^{mk}\Gamma_m(k)^{-1} \sum_{h \in C_m} (\det h)^{k-\kappa} e_m(hz) \tag{3.17}$$

(see, for example, the book of Maass [Maa]). Indeed, we notice that the only terms in the equality (3.17), correspond to $p = m, q = 0$ because of the poles of the Γ-functions in the denominator of (3.15). Also, the function $\omega(2\pi y, h; \alpha, \beta)$ reduces to the exponent $e_m(iyh)$ in view of the formulas:

$$\begin{aligned} &\xi(y, h; \alpha, 0) = \\ &i^{-m\alpha}2^{(1-\kappa)m}(2\pi)^{m\alpha}\Gamma_m(\alpha)^{-1}(\det h)^{\alpha-\kappa}e_m(iyh), \end{aligned} \tag{3.18}$$

$$\begin{aligned} &\xi(y, 0; \alpha, \beta) = \\ &i^{m\beta-m\alpha}2^{m(\kappa+1-\alpha-\beta)}\pi^{m\kappa}\Gamma_m(\alpha+\beta-\kappa)\Gamma_m(\alpha)^{-1}\Gamma_m(\beta)^{-1} \end{aligned} \tag{3.19}$$

$$\lim_{\xi \to 0} \xi(y, h; \kappa+s, s) = i^{-m\kappa}2^{\theta}\pi^{m\kappa}\Gamma_m(\kappa)^{-1}e_m(iyh), \tag{3.20}$$

with $q = 0$ and $\theta = [(m+p)/2]$. The formulas (3.18) - (3.20) are easily deduced from [Shi8], (1.31), (4.35.K); see also [Shi10], (7.11) - (7.14).

The confluent hypergeometric function $\omega(2\pi y, h; \alpha, \beta)$ can be used for analytic continuation of the Siegel – Eisenstein series [K1], [Fe], [Shi10] by means of the term by term analytic continuation of their Fourier coefficients. This coefficients can be expressed in terms of the functions (3.12) (see theorem 3.6 below). We list also some other properties of these functions, which are useful for the analytic continuation (see [Shi8], theorem 4.2):

functional equation

$$\omega(2\pi y, h; \alpha, \beta) = \omega(2\pi y, h; \kappa + (t/2) - \beta, \kappa + (t/2) - \alpha); \tag{3.21}$$

a uniform upper bound on compact subsets

$$|\omega(2\pi y, h; \alpha, \beta)| \le C_1 e^{-\tau(hy)}(1 + \mu(hy)^{-C_2}), \tag{3.22}$$

with α, β varying in a fixed compact subset $T \subset \mathbf{C}^2$ and the constants C_1, C_2 depending only on $T, \tau(x)$ being *the sum of eigenvalues* of a matrix x, μ *the minimum of their absolute values.*

3.3. Critical values of the confluent hypergeometric function. Now we give formulas, which express the function $\omega(2\pi y, h; \alpha, \beta)$ in terms of certain polynomials of the entries of the matrix $y = (y_{ij})$ provided $h > 0$ and either $\alpha - \kappa \in \mathbf{Z}, \alpha - \kappa \ge 0$ or $\beta \in \mathbf{Z}, \beta \le 0$ $(\kappa = (m+1)/2)$. The pairs (α, β) satisfying these conditions will be called *critical*: as we shall see, the critical values of s for the standard zeta function correspond to certain critical pairs. The following calculation of the special values is based on properties of the function $\zeta(z;\alpha,\beta)$ defined for $z \in \mathfrak{H}'_m = \{z \in \mathrm{M}_m(\mathbf{C}) | iz \in \mathfrak{H}_m\}$ by the integral

$$\zeta(z;\alpha,\beta) = \int_Y e^{-\mathrm{tr}(zx)} \det(x+1_m)^{\alpha-\kappa} \det x^{\beta-\kappa} dx, \tag{3.23}$$

which is absolutely convergent for Re $> \kappa - 1$ and defines a holomorphic function of (z,α,β). Let

$$\omega(z;\alpha,\beta) = \Gamma_m(\beta)^{-1} \det(z)^{\beta} \zeta(z;\alpha,\beta). \tag{3.24}$$

It was established by Shimura ([Shi8], theorem 3.1) that the function (3.24) can be analytically continued to a holomorphic function over $\mathfrak{H}'_m \times \mathbf{C}^2$ satisfying the functional equation

$$\omega(z;\kappa-\beta,\kappa-\alpha) = \omega(z;\alpha,\beta). \tag{3.25}$$

For an arbitrary compact subset $T \subset \mathbf{C}^2$ there exsist positive constants $A, B, > 0$ depending only on T such that

$$|\omega(z;\alpha,\beta)| \le A(1+\mu(y)^{-B}) \text{ for } y \in Y \subset \mathfrak{H}'_m, \quad (\alpha,\beta) \subset T. \tag{3.26}$$

It is known also (see [Shi8], (4.19)) that

$$\omega(y, 1_m;\alpha,\beta) = 2^{m(m+1)/2} e^{-\mathrm{tr}(y)} \omega(2y;\alpha,\beta) \tag{3.27}$$

and that for all $a \in \mathrm{GL}_m(\mathbf{R})$ one has

$$\omega({}^t a^{-1} y a^{-1};\alpha,\beta) = \omega(y;\alpha,\beta), \tag{3.28}$$

$$\omega(y,-h;\alpha,\beta) = \omega(y,h;\beta,\alpha) \tag{3.29}$$

$$\omega(y,h;\alpha,\beta) = 1. \tag{3.30}$$

Comparison of (3.27) and (3.28) shows that for $h > 0$ there is the identity

$$\omega(y,h;\alpha,\beta) = \omega(a(hy)a^{-1}, 1_m;\alpha,\beta) = 2^{m(m+1)/2} e^{-\mathrm{tr}(y)} \omega(2a(hy)a^{-1};\alpha,\beta), \tag{3.31}$$

This is done by taking a matrix $a \in \mathrm{GL}_m(\mathbf{R})$ with the condition $ah^t a = 1_m$.

Now let us consider the differential operator Δ_m (the *Maass differential operator*) acting on on C^∞-functions over $V \otimes \mathbf{C}$ of degree m, which is defined by the equality:

$$\Delta_m = \det(\partial_{ij}), \quad \partial_{ij} = 2^{-1}(1+\delta_{ij})\partial/\partial_{ij}. \tag{3.32}$$

For an integer $n \ge 0$ and a complex number β consider the polynomial

$$R(z;n,\beta) = (-1)^{mn} e^{\mathrm{tr}(z)} \det(z)^{n+\beta} \Delta_m^n [e^{\mathrm{tr}(z)} \det(z)^{-\beta}], \tag{3.33}$$

with $z \in V \otimes \mathbf{C}$, where the exponentiation is well defined by $\det(y)^\beta = \exp(\log(\det(y)))$ for $\det(y) > 0, y \in Y \otimes \mathbf{C}$. According to definition (3.33) the degree of the polynomial $R(z;n,\beta)$is equal to mn and the term of the highest degree coincides with $\det z^n$. We have also that for $\beta \in \mathbf{Q}$ the polynomial $R(z;n,\beta)$ has rational coefficients.

Proposition. *(See [Shi8], proposition 3.2). For any non negative integer n the functions $\det(z)^n\omega(z;n+\kappa,\beta)$ and $\det(z)^n\omega(z;\alpha,-n)$ are polynomial functions of z. More precisely, we have that*

$$\omega(z;n+\kappa,\beta)=\det{}^{-n}R(z;n,\beta), \tag{3.34}$$

$$\omega(z;\alpha,-n)=\omega(z;n+\kappa,n-\alpha)=\det{}^{-n}R(z;n,\kappa-\alpha). \tag{3.34a}$$

The proof is carried out using the following differentiation rule:

$$(-1)^{mn}\Delta_m^n\{e^{-\mathrm{tr}(z)}\det(z)^{-\beta}\omega(z;\alpha,\beta)\}=e^{-\mathrm{tr}(z)}\det(z)^{-\beta}\omega(z;\alpha+n,\beta), \tag{3.35}$$

which follows immediately from the definition (3.23), (3.24) and provides also an analytic continuation of the function $\omega(z;\alpha,\beta)$. The formulas (3.34) and (3.34a) follow then from the identities (3.35) and

$$\omega(z;\kappa,\beta)=\omega(z;\alpha,0)=1,$$

Notice also that for $m=1$ one has

$$R_1(z,n,\beta)=\sum_{k=0}^{n}\binom{n}{k}\beta(\beta+1)\cdots(\beta+k-1)z^{n-k}.$$

In §4, theorem 4.8, we establish the following more explicit expression for the function $R_m(z;n,\beta)$ of arbitrary degree m (see (4.32))

$$R_m(z;n,\beta)=$$
$$\sum_{r_1,\cdots,r_n=0}^{m} c_{m-r_1}(-\beta)c_{m-r_2}(-\beta-1)\cdots c_{m-r_n}(-\beta-n+1)\lambda_{r_1}(z)\cdots\lambda_{r_n}(z), \tag{3.36}$$

where

$$c_r(z)=\prod_{k=0}^{r-1}(\alpha+(k/2)),$$

and $\lambda_r(z)$ are polynomial functions of entries of the matrix variable $z\in \mathrm{M}_m(\mathbf{C})$ defined by

$$\det(t1_m-X)=\sum_{r=0}^{m}(-1)^r\lambda_r(X)t^{m-r} \tag{3.37}$$

In other words $\lambda_r(z)$ is the sum of all diagonal minors of size $r\times r$ of the matrix z.

If we apply this to functions $\omega(2\pi y,h;\alpha,\beta)$ from 3.2 then we get for $h>0$ the following identity:

$$\begin{aligned}&\omega(2\pi y,h;n+\kappa,\beta)=\omega(2\pi y,h;\kappa-\beta,n)=\\&2^{-m(m+1)/2}e_m(ihy)\omega(4\pi a^{-1}(hy)a\,;n+\kappa,\beta)=\\&2^{-m(m+1)/2}e_m(ihy)\det(4\pi hy)^{-n}R_m(4\pi hy\,;n,\beta).\end{aligned} \tag{3.38}$$

3.4. Proposition. *(Fourier expansion of the Siegel – Eisenstein series $E^*(z,s)$, see [Fe], §10). The series defined by (3.5) have the following Fourier expansion*

$$E^*(z,s) = \sum_{h \in N^{-1}A_m} b(h,y,s)e_m(hz), \tag{3.39}$$

in which coefficients have the form of the product

$$b(h,y,s) = N^{-m\kappa}W(y,h,s)\Gamma(h,s)RL^*(h,\chi,k+2s)M(h,\chi,k+2s), \tag{3.40}$$

with the factors described as follows ((a) - (d)):

(a)The confluent hypergeometric function

$$\begin{aligned}W(y,h,s) = i^{-mk}2^{\tau}\pi^{\sigma}\omega(2\pi y,h;k+s,s)\times \\ (\det y)^{\kappa-k-s}\delta_+(hy)^{k+s-\kappa+q/4}\delta_-(hy)^{s-\kappa+p/4},\end{aligned} \tag{3.41}$$

with (compare with (3.15))

$$\begin{aligned}\tau =&(2p-m)(k+s)+(2q-m)s+m+(m-r)+pq/2 = \\ &2(r-m)s+(2p-m)k+m+(m-r)(m+1)/2+pq/2,\\ \sigma =&p(k+s)+qs+m-r+\{(m-r)(m-r-1)-pq\}/2 = \\ =&rs+pk+\{(m-r)(m-r-1)-pq\}/2;\end{aligned}$$

(b) Gamma factor $\Gamma(h,s)$. Let for the integer r the symbol $\varepsilon(r)$ denote its parity: $\varepsilon(r)=0$, 1 with $r\equiv\varepsilon(r)$ mod 2. Put $\delta=\varepsilon(k), \mu=\varepsilon((r/2)+q+k)$ and then define: for $\varepsilon(r)=0$

$$\Gamma(h,s) = $$

$$= \frac{\Gamma_{m-r}\left(k+2s-\frac{m+1}{2}\right)\Gamma\left(s+\frac{k+\delta}{2}\right)\prod_{i=1}^{[m/2]}\Gamma(k+2s-i)}{\Gamma_{m-q}(k+s)\Gamma_{m-p}(s)\Gamma\left(s+\frac{k-m+r/2+\mu}{2}\right)\prod_{i=1}^{[(m-r)/2]}\Gamma(k+2s-m+i+(r-1)/2)},$$

and for $\varepsilon(r)=1$

$$\Gamma(h,s) = $$

$$= \frac{\Gamma_{m-r}\left(k+2s-\frac{m+1}{2}\right)\Gamma\left(s+\frac{k+\delta}{2}\right)\prod_{i=1}^{[m/2]}\Gamma(k+2s-i)}{\Gamma_{m-q}(k+s)\Gamma_{m-p}(s)\prod_{i=1}^{[(m-r-1)/2]}\Gamma(k+2s-m+i+r/2)},$$

(c) The ratio of Dirichlet L-functions RL^*. Let for a Dirichlet character χ modulo N of a parity $\delta = 0$ or 1

$$L_N^*(s,\chi) = \Gamma((k+\delta)/2)L_N(s,\chi) = \Gamma((k+\delta)/2)\prod_{q \nmid N}(1-\chi(q)q^{-s})^{-1} \tag{3.42}$$

denote the normalized Dirichlet L-function, which is regular for all $s \in \mathbf{C}, s \neq 1$, including $s=0$ (due to the condition $N>1$). Next we define an additional quadratic

Dirichlet character χ_h depending on $h \in A_m$ and defined only for even $r \neq 0$. Namely, for $h = 0$ let $\chi_h = \chi_0$ be trivial; for $h \neq 0$ we know that for some matrix $u \in \mathrm{GL}_m(\mathbf{Q})$

$$ {}^t uhu = \begin{pmatrix} h_1 & 0 \\ 0 & 0 \end{pmatrix} \quad \text{with } \det h_1 \neq 0, \tag{3.43} $$

then let χ_h denote the quadratic character attached to the quadratic field $\mathbf{Q}(\sqrt{\det h_1})/\mathbf{Q}$ (this definition does not depend on choice of a matrix u). Under these notation we set:

for an even r (i.e. for $\varepsilon(r) = 0$)

$$ RL^*(h, \chi, k + 2s) = $$
$$ \frac{L_N^*(k + 2s - m + (r/2), \chi\chi_h) \prod_{i=1}^{[(m-r)/2]} L_N^*(2k + 4s - 2m + r - 1 + 2i, \chi^2)}{L_N^*(k + 2s, \chi) \prod_{i=1}^{[m/2]} L_N^*(2k + 4s - 2i, \chi^2)}, $$

and for an odd r (i.e. with $\varepsilon(r) = 1$)

$$ RL^*(h, \chi, k + 2s) = $$
$$ \frac{\prod_{i=1}^{[(m-r-1)/2]} L_N^*(2k + 4s - 2m + r - 1 + 2i, \chi^2)}{L_N^*(k + 2s, \chi) \prod_{i=1}^{[m/2]} L_N^*(2k + 4s - 2i, \chi^2)}. $$

(d) The integral factor

$$ M(h, \chi, k + 2s) = \prod_{q \in P(h)} M_q(h, \chi(q) q^{-k-2s}) \tag{3.44} $$

is a finite Euler product, extended over primes q in the set $P(h)$ of prime divisors of the number N and of all elementary divisors of the matrix h. The important property of the product is that for each q we have that $M_q(h, t) \in \mathbf{Z}[t]$ is a polynomial with integral coefficients. The explicit form of it is insignificant for our purposes; however, one can find interesting explicit formulas for these coefficients in [Rag3], [Ki2].

3.5. Normalized Siegel – Eisenstein series. We introduce here three types of normalized Siegel – Eisenstein series in order to give a precise statement on their holomorphy properties with respect to the variable s, the properties of positivity of matrices ξ by which their Fourier coefficients are indexed, and also algebraic properties of these Fourier coefficients:

$$ G^*(z, s) = G^*(z.s; k, \chi, N) = $$
$$ N^{m(k+2s)/2} i^{mk} 2^{-m(k+1)} \pi^{-m(s+k)} \Gamma(1_m, s)^{-1} \times $$
$$ \times L_N^*(k + 2s, \chi) \prod_{i=1}^{[m/2]} L_N^*(2k + 4s - 2i, \chi^2) E(-(Nz)^{-1}, s) \det(\sqrt{N} z)^{-k} = \tag{3.45} $$
$$ N^{m(k+2s)/2} \tilde{\Gamma}(k, s) L_N(k + 2s, \chi) \prod_{i=1}^{[m/2]} L_N(2k + 4s - 2i, \chi^2) E^*(Nz, s), $$

with

$$E^*(Nz,s) = E(-(Nz)^{-1},s)\det(Nz)^{-k} = N^{-km/2}E|W(N),$$

$$\tilde{\Gamma}(k,s) =$$

$$i^{mk}2^{-m(k+1)}\pi^{-m(s+k)}\Gamma(1_m,s)^{-1}\Gamma((k+2s+\delta)/2)\prod_{j=1}^{[m/2]}\Gamma(k+2s-j) = \tag{3.46}$$

$$i^{mk}2^{-m(k+1)}\pi^{-m(s+k)} \times \begin{cases} \Gamma_m(k+s)\Gamma(s+(k-\frac{m}{2}+\mu)/2), & \text{if } m \text{ is even;} \\ \Gamma_m(k+s), & \text{otherwise.} \end{cases}$$

If m is odd then we put $G^+(z,s) = G^-(z,s) = G^*(z,s)$. If m is even then we define (with $\mu = \varepsilon((m/2)+k)$)

$$G^-(z,s) = \Gamma((k+2s-(m/2)+\mu)/2)^{-1}G^*(z,s), \tag{3.47}$$

$$G^+(z,s) = \frac{i^\mu \pi^{(1/2)-k-2s+(m/2)}}{\Gamma((1-k-2s+(m/2)+\mu)/2)}G^*(z,s) = \tag{3.48}$$

$$\frac{2i^\mu\Gamma(k+2s-(m/2))\cos(\pi(k+2s-(m/2)-\mu)/2)}{(2\pi)^{k+2s-(m/2)}}G^-(z,s).$$

We shall see in §3 of chapter 3 that the normalizing factors in formulas (3.45), (3.47) and (3.48) are closely related to those of the Dirichlet L-series and the standard zeta functions (for even m).

Now we reformulate proposition 3.4 for the normalized series $G^*(z,s)$

$$G^*(z,s) = \sum_{h\in A_m} b^*(h,y,s)e_m(hz), \tag{3.49}$$

where

$$b^*(h,y,s) = W^*(y,h,s)\Gamma^*(h,s)L_N^*(h,\chi,k+2s)M(h,\chi,k+2s),$$

with

$$\Gamma^*(h,s) = \Gamma^-(1_m,s)^{-1}\Gamma^-(h,s),$$

$$W^*(y,h.s) = N^{m(s+k-\kappa)}i^{mk}2^{-m(k+1)}\pi^{-m(s+k)}e_m(-ihy)W(Ny,N^{-1}h,s) =$$

$$i^{mk}2^{-m(k+1)}\pi^{-m(s+k)}e_m(-ihy)W(y,h,s).$$

The factor $M(h,\chi,k+2s)$ is given by (3.44), and for r is even we have that

$$L_N^*(h,\chi,k+2s) = L_N^*(k+2s-m+(r/2),\chi\chi_h)\prod_{i=1}^{[(m-r)/2]}L_N^*(2k+4s-2m+r-1+2i,\chi^2)$$

and for r odd

$$L_N^*(h,\chi,k+2s) = \prod_{i=1}^{[(m-r-1)/2]} L_N^*(2k+4s-2m+r-1+2i,\chi^2)$$

3.6. Theorem *(On holomorphy properties of Fourier coefficients of the Siegel – Eisenstein series, see [Fe], theorem 9.1.) Let $2|N, N>1$. Then*
(a) If $\chi^2 \neq 1$ then the function $G^*(z,s)$ is an entire function of the variable s;
(b) Suppose that χ^2 is trivial, then we have

(b_1) if either $2k \geq m$ and m odd or $2k \geq m$ and m is even, but $(m/2)+k$ is odd (i.e. $\mu = 1$), then the function $G^*(z,s)$ is an entire function of the variable s;

(b_2) if $2k \geq m$ and both numbers m and $(m/2)+k$ are even (i.e. $\mu = \varepsilon(m) = 0$) then the function $G^*(z,s)$ is an entire function of the variable s with the exclusion of a possible simple pole at the point $s = (m+2-2k)/4$;

(b_3) if $m > 2k \geq 0$ then the function $G^*(z,s)$ is an entire function of the variable s with possible exclusion of simple poles at those points s for which $2s$ is an integer and $[(m-2k+3)/2] \leq 2s \leq (m+1-2k)/2$;

(b_4) if $k=0$ then the function $G^*(z,s)$ has a simple pole at the point $s=(m+1)/2$ iff χ is trivial, and in this case we have that the function

$$\mathrm{Res}_{s=(m+1)/2} G^*(z,s;0,1,N)$$

of the variable z is a non-zero constant.

3.7. Theorem. *(On positivity properties of Fourier expansions of the normalized Siegel – Eisenstein series). Assume that $2k > m$ and define the numbers $A(\chi)$, $B(\chi)$, $C(\chi,k)$ as follows:*

(a) If χ^2 is not trivial, m is even and $\mu = \varepsilon(k+(m/2))$ then put

$$A(\chi) = B(\chi) = 1+(m/2), \quad C(\chi,k) = (m-2k+2-2\mu)/4;$$

(b) If χ^2 is trivial, m is even and $\mu = \varepsilon(k+(m/2))$ then put

$$A(\chi) = B(\chi) = (m/2), \quad C(\chi,k) = (m-2k+2-2\mu)/4;$$

(c) If χ^2 is not trivial, m is odd,then put

$$A(\chi) = B(\chi) = (m+3)/2, \quad C(\chi,k) = [(1+m-2k)/4];$$

(d) If χ^2 is trivial, m is odd,then put

$$A(\chi) = (m+5)/2, \quad B(\chi) = (m+1)/2, \quad C(\chi,k) = [(3+m-2k)/4];$$

under these notation and assumptions there are the following positivity properties of matrices $h \in A_m$, by which the Fourier coefficients of the series $G^*(z,s)$ are indexed:

1) if $s \leq 0, s \in \mathbf{Z}$ and $k+2s \geq A(\chi)$ then

$$G^*(z,s) = \sum_{A_m \ni h > 0} b^*(h,y,s)e_m(hz), \tag{3.50}$$

2) if $k+s-\kappa\in\mathbf{Z}, k+s-\kappa\geq 0$ $(\kappa=(m+1)/2), s\leq C(\chi,k)$ then

$$G^*(z,s)=\sum_{A_m\ni h\geq 0} b^*(h,y,s)e_m(hz), \tag{3.51}$$

The proof of the theorem 3.7 is contained in the book of P.Feit [Fe], theorems 14.1.A – 14.1.F; it is based on a detailed study of poles and residues of the Γ-factor $\Gamma^*(h,s)$ and of the Dirichlet L-function $L^*(h,\chi,k+2s)$, which was carried out in [Fe] in terms of the functions $f(n,s)=\Gamma(n+s)/\Gamma(s)$; for positive integers n these functions turn out to be polynomials with zeros given by $s\in\mathbf{Z}, s\leq 0, s+n>0$. It was shown in [Fe] that the factor $\Gamma^*(g,s)=\Gamma(h,s)/\Gamma(1_m,s)$ is equivalent (up to multiplication by an invertible entire function) to a certain explicitly given polynomial in $\mathbf{C}[s]$ (see [Fe], §11). It follows also from this calculation that

(a) if χ^2 is trivial, m is odd and $s=s_0=(m+2-2k)/4$ then the function $G^*(z,s)$ has a pole at the point $s=s_0$ such that the residue

$$\mathrm{Res}_{s=s_0}G^*(z,s) \text{ has a non negative Fourier expansion} \tag{3.52}$$

(theorem 14.1.C);

(b) if χ^2 is trivial and m is odd then the function $G^*(z,s)$ is finite at the point $s=s_0$ and has a non negative Fourier expansion.

It is essential for our purposes to reformulate the corresponding statements for the series $G^+(z,s)$ and $G^-(z,s)$ (see(3.47), (3.48)), which are obtained from $G^*(z,s)$ by an additional normalization. The following theorem is an immediate consequence of the theorem 3.6 on holomorphy and the properties (3.50)-(3.52).

3.8. Theorem *(on Fourier coefficients with positive matrix numbers). Let m is even, $2k>m$. Then:*

(a) For $2s$ to be an integer, $s\leq 0, k+2s\geq 1+(m/2)$ there is the following Fourier expansion

$$G^+(z,s)=\sum_{A_m\ni h>0} b^+(h,y,s)e_m(hz), \tag{3.53}$$

where for $s>(m+2-2k)/4$ in (3.54) non-zero terms only occur for positive definite $h>0$, and for all s from (a) with $h>0, h\in A_m$ the following identity holds

$$b^+(h,y,s)=W^*(y,h,s)L_N^+(k+2s-(m/2),\chi\chi_h)M(h,\chi,k+2s),$$

with

$$L^+(s,\chi)=\frac{2i^\delta\Gamma(s)\cos(\pi(s-\delta)/2)}{(2\pi)^s}L_N(s,\chi)$$

is the normalized Dirichlet L-function, $\delta=0$ or 1 according to $\chi(-1)=(-1)^\delta$, the factor $M(h,\chi,k+2s)$ is defined by (3.44),

$$W^*(y,h,s)=2^{-m\kappa}\det h^{k+2s-\kappa}\det(4\pi y)^s R(4\pi hy;-s;\kappa-k-s),$$

provided s is an integer, where $R(y;n,\beta)$ defined by (3.33), and $b^+(h,y,s)=0$ otherwise (if $s\notin\mathbf{Z}$).

(b)If $2s$ is an integer with $k+2s \leq m/2, k+s \geq \kappa$ then there is the following Fourier expansion

$$G^-(z,s) = \sum_{A_m \ni h \geq 0} b^-(h,y,s)e_m(hz), \tag{3.53a}$$

and for all s from (b) with $h > 0, h \in A_m$ the following identity holds

$$b^-(h,y,s) = W^*(y,h,s)L_N^-(k+2s-(m/2), \chi\chi_h)M(h,\chi,k+2s),$$

where

$$L^+(s,\chi) = L_N(s,\chi),$$
$$W^*(y,h,s) = 2^{-m\kappa}\det(4\pi y)^{\kappa-k-s}R(4\pi hy; -s; k+s-\kappa, s),$$

provided $s+k-\kappa$ is an integer, and $b^-(h,y,s)=0$ otherwise.

The proof is deduced from the expansions (3.49) if we remember the definition of the normalizing factors and the positivity property in the theorem 3.7. We also note that by (3.41)

$$W^*(y,h,s) =$$
$$e_m(-ihy)\omega(2\pi y, h; k+s, s)(\det y)^{\kappa-k-s}\delta_+(hy)^{k+s-\kappa+q/4}\delta_-(hy)^{s-\kappa+p/4},$$

and then take into account the formula (3.38) for the critical values of the function ω. In case of the odd parity $2s \in \mathbf{Z}$ we get vanishing of the Fourier coefficients because of the Γ-factors in (3.47), (3.48).

§4. Holomorphic projection operator and the Maass differential operator

4.1. Holomorphic projection operator. We start with describing a vector space on which this operator acts. A function

$$F : \mathfrak{H}_m \to \mathbf{C}, \quad F \in C^\infty(\mathfrak{H}_m)$$

is called a C^∞*-modular form* of weight k on the group $\Gamma_0^m(N)$ with a Dirichlet character ψ mod N if the following automorphy condition is satisfied:

$$F((az+b)(cz+d)^{-1}) = \psi(\det d)\det(cz+d)^k F(z)$$

for all

$$\gamma \in \begin{pmatrix} a & b \\ c & d \end{pmatrix} \in \Gamma_0^m(N)$$

(compare with §1, (1.11)). The space of functions F with the above condition is denoted by $\tilde{\mathcal{M}}_m^k(N,\psi)$. For all $F \in \tilde{\mathcal{M}}_m^k(N,\psi)$ there is the following Fourier expansion

$$F(z) = \sum_{h \in A_m} A(y,h)e_m(hx), \tag{4.1}$$

where $A(y,h)$ are some C^∞-functions on Y. The Petersson inner product is defined for an arbitrary holomorphic cusp form $f \in \mathcal{S}_m^k(N,\psi)$ and $F \in \tilde{\mathcal{M}}_m^k(N,\psi)$ by

$$\langle f, F\rangle_N = \int_{\Phi_0(N)} \overline{f(z)}F(z)\det y^{k-m-1}dx\,dy,$$

where $\Phi_0(N) = \Gamma_0^m(N)\backslash\mathfrak{H}_m$ is a fundamental domain for the group $\Gamma_0^m(N)$.

We call a function $F \in \tilde{\mathcal{M}}_m^k(N,\psi)$ *a function of a bounded growth* if for each $\varepsilon > 0$ the following integral converges:

$$\int_X \int_Y |F(z)|\det y^{k-1-m} e^{-\varepsilon \mathrm{tr}(y)} dy\, dx \; < \; \infty \tag{4.2}$$

where

$$X = \{x \in M_m(\mathbf{R}) | {}^t x = x, |x_{ij}| \le 1/2 \text{ for all } i,j\},$$
$$Y = \{y \in M_m(\mathbf{R}) | {}^t y = y > 0\}.$$

Respectively, we call a function $F \in \tilde{\mathcal{M}}_m^k(N,\psi)$ *a function of a moderate growth* if for all $z \in \mathfrak{H}_m$ and for all sufficiently large values of $\mathrm{Re}(s) \gg 0$ the integral

$$\int_{\mathfrak{H}_m} F(w)\det(\overline{w} - z)^{-k-|2s|}\det\mathrm{Im}(w)^{k+s} d^\times w \tag{4.3}$$

is absolutely convergent and admits an analytic continuation over s to the point $s = 0$. The last definition may differ from a traditional one; its meaning is clarified by the following result (theorem 4.2), which provides a refinement of theorem 1 of the Sturm's paper [St2]. It will follow from the proof that all functions of bownded growth automatically turn out to be of a moderate growth in the sense of definitions (4.2), (4.3) given above.

4.2. Theorem. *Let $F \in \tilde{\mathcal{M}}_m^k(N,\psi)$ and $k > 2m$. Put for $h > 0, h \in A_m$*

$$a(h) = c(k,m)^{-1}\det(4h)^{k-(m+1)/2} \int_Y A(y,h)e_m(ihy)\det y^{k-1-m}\, dy, \tag{4.4}$$

with

$$c(t,m) = \Gamma_m(t-(m+1)/2)\pi^{-m(t-(m+1)/2)},$$

and $A(y,h)$ being coefficients of the expansion (4.1) and suppose that the integral (4.4) is absolutely convergent. Define the function

$$\mathcal{H}ol\, F(z) = \sum_{A_m \ni h > 0} a(h)e_m(hz). \tag{4.5}$$

Then

(a) if the function $F \in \tilde{\mathcal{M}}_m^k(N,\psi)$ is of a bounded growth then $\mathcal{H}ol\, F(z) \in \mathcal{S}_m^k(N,\psi)$.

(b) If the function $F \in \tilde{\mathcal{M}}_m^k(N,\psi)$ is of a moderate growth and the expansion (4.1) contains only terms with positive definite matrices $h \in A_m$, then $\mathcal{H}ol\, F(z) \in \mathcal{M}_m^k(N,\psi)$.

In both cases for all $g \in \mathcal{S}_m^k(N,\psi)$ the following equality holds:

$$\langle g, F\rangle_N \;=\; \langle g, \mathcal{H}ol\, F\rangle_N. \tag{4.6}$$

Remark. *The cusp form $\mathcal{H}ol\, F$ is uniquely determined by (4.6) under the assumptions of (a), but in (b) this equality is not sufficient to identify the modular form $\mathcal{H}ol\, F$. For example, (4.6) does not change if we replace this modular form by adding to it an*

Eisenstein series (of Siegel or of Klingen type). the part (a) of the theorem 4.2 was established by Sturm [St2].

4.3. Poincaré series of two variables (of exponential type) of higher level. In order to prove the theorem 4.2 we use this kind of Poincaré series introduced by Klingen [Kl3] and used by Böcherer in [Bö4] instead of Poincaré series of one variable from [St2], [Gr-Za]. We consider an element $(\Gamma g\Gamma)$ of the Hecke algebra $\mathcal{L}^m(N)$ with $\Gamma = \Gamma_0^m(N)$,

$$g \in \Delta = \Delta_q^m =$$
$$\{\alpha = \begin{pmatrix} a & b \\ c & d \end{pmatrix} \in G_{\mathbf{Q}+} \cap \mathrm{GL}_2m(\mathbf{Z}[q^{-1}]) | \nu(\alpha)^{\pm} \in \mathbf{Z}[q^{\pm}], c \equiv 0_m(\mathrm{mod} N)\},$$

where ${}^t g J_m g = \nu(g) J_m,\ \nu(g) > 0$.Put

$$P_m^k(z, w, g, s) = \sum_{g \in \Gamma g \Gamma} \psi(\det a) j(\gamma, z)^{-k-|2s|} \det(\gamma(z) + w)^{k-|2s|}, \tag{4.7}$$

$\gamma = \begin{pmatrix} a & b \\ c & d \end{pmatrix} \in \Gamma g \Gamma$ and $u^{k-|2s|} \overset{\mathrm{def}}{=} u^{-k}|u|^{-2s}$ for $u \in \mathbf{C}^{\times}, s \in \mathbf{C}$. The series in (4.7) converges absolutely and uniformly on products of the type $V_m(d) \times V_m(d)$ for $k + \mathrm{Re}(2s) > 2m + 1,\ d > 0$,

$$V_m(d) = \left\{ z = x + iy \in \mathfrak{H}_m | y \geq d1_m, \mathrm{tr}({}^t x\, x) \leq \frac{1}{d} \right\}.$$

We also put

$$\mathcal{P}_m^k(z, w, g, s) = \det \mathrm{Im} z^s \det \mathrm{Im}\, w^s P_m^k(z, w, g, s). \tag{4.8}$$

The following properties of these series were established by Böcherer in [Bö4]:

(a) symmetry

$$\mathcal{P}_m^k(z, w, g, s) = \mathcal{P}_m^k(w, z, g, s) \tag{4.9}$$

(b) automorphy with respect to both arguments:

$$\begin{gathered} \mathcal{P}_m^k(\gamma(z), \gamma(w), g, s) = \\ \psi(\det d_1)\psi(\det d_2) j(\gamma_1, z)^k (\gamma_2, w)^k \mathcal{P}_m^k(z, w, g, s), \end{gathered} \tag{4.10}$$

where $\gamma_i = \begin{pmatrix} a_i & b_i \\ c_i & d_i \end{pmatrix} \in \Gamma,\ i = 1, 2;$

(c) action of Hecke operators:

$$\begin{gathered} \mathcal{P}_m^k(z, 1_m, g, s)|_{k,\psi}(\Gamma g \Gamma)_z = \mathcal{P}_m^k(z, w, g, s) = \\ \mathcal{P}_m^k(z, w, 1_m, s)|_{k,\psi}(\Gamma g \Gamma)_w, \end{gathered} \tag{4.11}$$

where the subscript indicates to which of the variables the Hecke operator is being applied with the action defined by (1.21).

(d) the integral representation: for all $f \in \mathcal{S}_m^k(N, \psi)$ we have that

$$\langle \mathcal{P}_m^k(-\overline{z}, w, g, s), f(w) \rangle_{N,w} = \mu(m, k, s) f|_{k,\psi}(\Gamma g \Gamma)(z), \tag{4.12}$$

with

$$\mu(m,k,s) = 2^{m+\frac{m(m+1)}{2}-2ms+1} i^{-mk} \pi^{\frac{m(m+1)}{2}} \frac{\Gamma_m(k+s-(m+1)/2)}{\Gamma_m(k+s)}.$$

The proof of the properties (4.9), (4.11) is easily deduced from the symmetry relation

$$j(\gamma,z)\det(\gamma(z)+w) = j(\tilde{\gamma},w)\det(\tilde{\gamma}(w)+z), \tag{4.13}$$

which is valid for all $\gamma \in \mathrm{Sp}_m(\mathbf{R})$ with

$$\tilde{\gamma} = \gamma\left[\begin{pmatrix} 1_m & 0_m \\ 0_m & -1_m \end{pmatrix}\right]$$

and $\Gamma g \Gamma = \Gamma \tilde{g} \Gamma$ for $g \in \Delta$.Then (4.10) is immediately deduced from the definition (4.7). Proof of the integral formula (4.12) is carried out similarly to that in Klingen's article [Kl5]; for this purpose we may assume that $g = 1_m$. The integration in the left hand side of (4.12) can be reduced by the standard unfolding procedure to that over the whole Siegel upper half plane: notice that

$$\mathfrak{H}_m = \cup_\gamma \gamma(\Phi_0(N)), \quad \gamma \in \Gamma = \Gamma_0^m(N),$$

where $\Phi_0(N) = \Gamma_0^m(N)\backslash\mathfrak{H}_m$ is a fundamental domain for the group $\Gamma_0^m(N)$. The required property follows then from the integral representation

$$\begin{gathered}\int_{\mathfrak{H}_m} f(w)\det(\overline{w}-z)^{-k-|2s|}\det \mathrm{Im}(w)^{k+s} d^\times w = \\ i^{mk} 2^{m(m+1)-2ms-mk} I_m(k+s-m-1) f(z) \det \mathrm{Im}(z),\end{gathered} \tag{4.14}$$

where

$$\begin{aligned} I_m(s) = \int_{E_m} & \det(1_m - \overline{w}w)^s du\, dv = \\ & \frac{\pi^{\frac{m(m+1)}{2}}}{2} \frac{\Gamma_m(s+1+(m-1)/2)}{\Gamma_m(s+m+1)} \end{aligned} \tag{4.15}$$

denotes the integral studied by Hua Lo-Ken [HLK]. This integral is absolutely convergent for $\mathrm{Re}(s) > -1$, $f \in \mathcal{S}_m^k(N,\psi)$, the integration in (4.14) is extended over the generalized unit disc

$$E_m = \{w = u + iv \in \mathrm{M}_m(\mathbf{C}) | {}^t w = w, 1_m - \overline{w}w > 0\}$$

of the degree m, the image of $\mathfrak{H}_m$ via the Cayley transform

$$w \;\mapsto\; (w - i1_m)(w + i1_m)^{-1} \quad (w \in \mathfrak{H}_m).$$

In order to prove (4.14) we note that if $f \in \mathcal{S}_m^k(N,\psi)$ then there is the following upper estimate

$$f(z) \le c \det \mathrm{Im}(w)^{-k/2}$$

and the integral in (4.14) is majorated by

$$\int_{\mathfrak{H}_m} |\det(\overline{w}-z)|^{-k-2\mathrm{Re}(s)} \det \mathrm{Im}(w)^{(k/2)+\mathrm{Re}(s)} d^{\times} w,$$

which provides the absolute convergence of (4.14) in this domain for $(k/2)+\mathrm{Re}(s) > m$. Next we rewrite the integrand in (4.14) in the form

$$g(w)\det(\overline{w}-z)^{-k-s} \det \mathrm{Im}(w)^{(k+s)}$$

with the holomorphic function

$$g(w) = f(w)\det(w-\overline{z})^{-s},$$

which is integrated using the Cayley transform, so that (4.14) follows.

4.4. Reduction of theorem 4.2 to properties of Poincaré series. We restrict ourselves to the case of functions F satisfying assumptions of (b). Set in formulas (4.10), (4.12) $g = 1_m$ and define a function of two variables $K_m^k(z, w, s)$ by the equality

$$K_m^k(z, w, s) = \mu(m, k, s)^{-1} \mathcal{P}_m^k(-\overline{z}, w, 1_{2m}, s). \tag{4.16}$$

We show that the function

$$\mathcal{H}ol\, F(z, s)|_{s=0} = \langle K_m^k(z, w, s), F(w)\rangle_{N,w}|_{s=0} \tag{4.17}$$

obtained by analytic continuation of the right hand side to the point $s = 0$ satisfies all conditions of the theorem, i.e. it coincides with the function defined by (4.4), (4.5): $(\mathcal{H}ol\, F)(z, s)|_{s=0} \in \mathcal{M}_m^k(N, \psi)$ and the equality

$$\langle g(z), F(z)\rangle_N \quad = \quad \langle g(z), (\mathcal{H}ol\, F)(z, s)|_{s=0}\rangle_N$$

holds for all $g \in \mathcal{S}_k^m(N, \psi)$. For this purpose we note that for sufficiently large value of $\mathrm{Re}(s)$ the right hand side of (4.17) can be rewritten in the form of an integral over the whole Siegel upper half plane $\mathfrak{H}_m$ of degree m:

$$\begin{aligned} &\langle K_m^k(z, w, s), F(w)\rangle_{N,w} = \\ &\mu(m, k, s)^{-1} \det \mathrm{Im}(z)^s \int_{\mathfrak{H}_m} F(w)\det(\overline{w}-z)^{-k-|2s|} \det \mathrm{Im}(w)^{k+s} d^{\times} w \end{aligned} \tag{4.18}$$

(due to the assumption on the growth of F the integral (4.18) is absolutely convergent for all $\mathrm{Re}(s) \gg 0$). Next let us consider the subgroup

$$\Gamma_\infty^0 = \left\{ \gamma = \pm \begin{pmatrix} 1_m & b \\ 0 & 1_m \end{pmatrix} | \gamma \in \Gamma \right\} \subset \Gamma = \Gamma_0^m(N).$$

Then the set

$$\{w = u + iv \in \mathfrak{H}_m | u \in X, v \in Y\}$$

is a fundamental domain for the action of Γ^0_∞ on $\mathfrak{H}_m$, and we see that for $\mathrm{Re}(s) \gg 0$ the right hand side of (4.18) takes the form

$$2\mu(m,k,s)^{-1}\det \mathrm{Im}(z)^s \int_X\int_Y F(w)\sum_{b\in L}\det(\overline{w}-z+b)^{-k-|2s|}\det \mathrm{Im}(w)^{k+s}d^\times w =$$

$$2\mu(m,k,s)^{-1}\det \mathrm{Im}(z)^s \int_X\int_Y F(w)S(\overline{w}-z,L;k+s,s)\det \mathrm{Im}(w)^{k+s}d^\times w, \tag{4.19}$$

where $L = \mathrm{M}_m(\mathbf{Z})\cap V$ is a lattice in $V = \{x \in \mathrm{M}_m(\mathbf{R})|{}^tx = x\}$, and the function

$$S(z,L;k+s,s) \quad = \quad \sum_{b\in L}\det(z+b)^{-k-|2s|}$$

admits an analytic continuation to all $s \in \mathbf{C}$ by means of the Fourier expansion of (4.14). Moreover, for $k > m$ we have that

$$S(\overline{w}-z,L;k+s,s)|_{s=0} = (-2\pi i)^{mk}\Gamma_m(k)^{-1}\sum_{A_m\ni h>0}\det h_{k-\kappa}e_m(f(z-\overline{w})) \tag{4.20}$$

Under the assumption on growth of F the integral admits an analytic continuation to the point $s = 0$. This analytic continuation can be explicitly given in the form of a Fourier expansion by means of (4.19), (4.20) using the positivity of h. As a result the function $\mathcal{H}ol\, F(z,s)|_{s=0}$ takes the form

$$\mathcal{H}ol\, F(z,s)|_{s=0} = (4\pi)^{m(k-(m+1)/2)}\Gamma_m(k-(m+1)/2)^{-1}\times$$
$$\sum_{A_m\ni h>0}\det h^{k-(m+1)/2}\int_X\int_Y F(z)e_m(h(z-\overline{w}))\det \mathrm{Im}w^k d^\times w.$$

Then the formula (4.4) follows from the obvious equality:

$$\int_X\int_Y F(z)e_m(h(z-\overline{w}))\det \mathrm{Im}w^k d^\times w =$$
$$= e_m(hz)\int_Y A(v,h)e_m(ihv)\det v^{k-1-m}dv.$$

In order to prove the remaining statements we note that the function $\mathcal{H}ol\, F(z,s)|_{s=0}$ defined by (4.17) is holomorphic and it satisfies the automorphic properties with respect to $\Gamma_0^m(N)$ of weight k with the Dirichlet character ψ mod N. Indeed, these properties are satisfied by the function $K^k_m(z,w,s)$ and consequently by (4.18) for $\mathrm{Re}(s) \gg 0$. But the identity expressing the automorphy condition (1.11) does not change by analytic continuation, and we get $\mathcal{H}ol\, F(z,s)|_{s=0} \in \mathcal{M}^m_k(N,\psi)$ (note for $m > 1$ also the Koecher principle is applicable). The equality (4.6) is then deduced from definitions (4.16) and (4.17), and from the automorphy property (4.12) of the Poincaré series. For $\mathrm{Re} \gg 0$ we get

$$\langle g, \langle K^k_m(z,w,s), F(w)\rangle_{N,w}\rangle_{N,z} = \langle g,\, F\rangle_N =$$
$$= \langle \overline{\langle K^k_m(z,w,s), g(z)\rangle_{N,z}}, F(z)\rangle_{N,w}. \tag{4.21}$$

These equalities remain valid by the analytic continuation and we get (4.6). In the equality (4.21) the property

$$\overline{K^k_m(z,w,s)} \quad = \quad K^k_m(z,w,\overline{s}).$$

was taken into account.

4.5. When applying the formula (4.4), it is convenient to use the integral representation (3.9) for the Γ-function of degree m,

$$\Gamma_m(s) = \pi^{m(m-1)/4} \prod_{j=0}^{m-1} \Gamma(s-(j/2)),$$

This integral representation can be rewritten in the equivalent form

$$\begin{aligned}\Gamma_m(\nu-(m+1)/2)\mathrm{det}u^{(m+1)/2-\nu} &= \int_Y (\mathrm{det}y)^{\nu-m-1}e^{-\mathrm{tr}(uy)}\,dy = \\ &= \int_Y (\mathrm{det}y)^{\nu-(m+1)/2}e_m(i(2\pi)^{-1}uy)\,d^\times y.\end{aligned} \tag{4.22}$$

Moreover, if $R(y) \in \mathbf{C}[y_{ij}]$ is a polynomial of $y = y_{ij}$, $i \leq j$ then for all $\nu \in \mathbf{Z}, \nu > m$ we have that

$$\begin{aligned}&\int_Y R(y)(\mathrm{det}y)^{\nu-(m+1)/2}e^{-\mathrm{tr}(uy)}\,d^\times y = \\ &\int_Y [R(\partial/\partial u)e^{-\mathrm{tr}(uy)}](\mathrm{det}y)^{\nu-(m+1)/2}\,d^\times y = \\ &R(\partial/\partial u)[\Gamma_m(\nu-(m+1)/2)\mathrm{det}u^{(m+1)/2-\nu}],\end{aligned} \tag{4.23}$$

where $\partial/\partial u \;=\; \partial_{ij}$, $\partial_{ij} = 2^{-1}(1+\delta_{ij}\partial/\partial u_{ij})$. Indeed, it suffices to verify the statement (4.23) for monomials of the form

$$R(y) = \prod_{i\leq j} y_{ij}^{a(i,j)}, \quad a(i,j) \in \mathbf{Z},\; a(i,j) \geq 0.$$

In this particular case this is done by application of the differential operator

$$\prod_{i\leq j} \partial/\partial u_{ij}^{a(i,j)}$$

to both sides of the equality (4.22). We shall need formulas (4.4) and (4.23) in a special situation, which is described in the next theorem below.

4.6. Theorem. *Let C^∞-modular form $F \in \mathcal{M}_k^m(N,\psi)$ has the form of a product of the type $F(z) = g(z)G(z)$, where*

$$g(z) = \sum_{A_m \ni h > 0} B(h) e_m(hz),$$

$$G(z) = \sum_{A_m \ni h \geq 0} C(h) \det(4\pi y)^{-n} R(4\pi hy; n, \beta) e_m(hz),$$

$F(z)$ satisfies one of the two assumptions (a) or (b) of the theorem 4.2 and $R(z;n,\beta)$ is the polynomial (3.33) defined for any integer $n \geq 0$, $\beta \in \mathbf{C}$ and $z = {}^t z \in \mathrm{M}_n(\mathbf{C})$ by

$$R(z;n,\beta) = (-1)^{mn} e^{\mathrm{tr}(z)} \det(z)^{n+\beta} \Delta_m^n [e^{\mathrm{tr}(z)} \det(z)^{-\beta}],$$

where

$$\Delta_m = \det(\partial_{ij}), \quad (\partial_{ij} = 2^{-1}(1+\delta_{ij})\partial/\partial_{ij} z, \quad i \leq j)$$

is the Maass differential operator. Then the following equality holds

$$\mathcal{H}ol\, F(z) = \sum_{A_m \ni h = h_1 + h_2 > 0} B(h_1) C(h_2) P(h_2, h; n, \beta) e_m(hz), \tag{4.24}$$

where $P(v,u) = P(v,u;n\beta)$ denotes a polynomial of $u = {}^t u = (u_{ij})$ and $v = {}^t v = (v_{ij})$ with the property

$$P(v,u;n,\beta) \equiv \det v^n (\mathrm{mod} \langle u_{ij} \rangle) \tag{4.25}$$

and $P(v,u;n,\beta) \in \mathbf{Q}[u,v]$ for any $\beta \in \mathbf{Q}$.

Proof of the theorem 4.6 is carried out by a straightforward application of the integral formula (4.4) for the action of $\mathcal{H}ol$ on each of the Fourier coefficients of the function $F(z)$:

$$A(y,h) = \sum_{A_m \ni h = h_1 + h_2 > 0} B(h_1) C(h_2) \det(4\pi y)^{-n} R(4\pi h_2 y; n, \beta) e_m(ihz).$$

As a result we get

$$A(h) = \sum_{A_m \ni h = h_1 + h_2 > 0} B(h_1) C(h_2) P(h_2, h; n, \beta),$$

where

$$P(v,u) = P(v,u;n,\beta) =$$

$$\frac{\det(4\pi u)^{k-(m+1)/2}}{\Gamma_m(k-(m+1)/2)} \int_Y R(4\pi vy;n,\beta)\det(4\pi y)^{-n}\det y^{k-(m+1)/2} e_m(2iuy)d^\times y =$$

$$\frac{\det(4\pi u)^{k-(m+1)/2}}{\Gamma_m(k-(m+1)/2)} \int_Y R(4\pi vy;n,\beta)\det(4\pi y)^{-n+k-(m+1)/2} e_m(2iuy)d^\times y = \tag{4.26}$$

$$\frac{\det(4\pi u)^{k-(m+1)/2}}{\Gamma_m(k-(m+1)/2)} \int_Y R(vy;n,\beta)\det(y)^{-n+k-(m+1)/2} e^{-\mathrm{tr} uy} d^\times y =$$

$$\frac{\Gamma_m(k-n-(m+1)/2)}{\Gamma_m(k-(m+1)/2)} \det u^{k-(m+1)/2} R(v\cdot\partial/\partial u;n,\beta)[\det u^{(m+1)/2-k+n}],$$

with $n \in \mathbf{Z}, n \geq 0, \beta \in \mathbf{C}$. In order to accomplish the proof it suffices to show that the function $P(v,u) = P(v,u;n,\beta)$ is a polynomial with the desired properties (4.25). This last fact is deduced from the last of the equalities (4.26) and some general properties of the differential operator $\partial/\partial u$ which are given below (see also [Kl5]).

4.7. Let us consider the natural representation

$$\rho_r : \mathrm{GL}_m(\mathbf{C}) \to \mathrm{GL}(\Lambda^r\mathbf{C}^m)$$

of the group $\mathrm{GL}_m(\mathbf{C})$ on the vector space $\Lambda^r\mathbf{C}^m$ and put

$$\rho_r^*(x) = \det(x)\rho_{r-m}{}^t x^{-1} \quad (r = 0,1,\cdots,m)$$

Then the representations ρ and ρ_r^* turn out to be polynomial representations so that for each $x \in \mathrm{M}_m(\mathbf{C})$ the linear operators $\rho_r(x)$, ρ_r^* are well defined. We consider the differential operators $\rho_r(\partial/\partial z)$ and $\rho_r^*(\partial/\partial z)$ which associate to each $\mathbf{C}$-valued function on $\mathfrak{H}_m$ a certain $\mathrm{M}_t(\mathbf{C})$ valued function on $\mathfrak{H}_m$ with $t = \binom{m}{r}$. In particular we put

$$\Delta = \rho_m(\partial/\partial z) = \rho_m^*(\partial/\partial z) = \det\left(2^{-1}(1+\delta_{ij})\partial/\partial_{ij}z\right).$$

The following differentiation rules are valid (see [Shi8], lemma 9.1):
(a) If f and g are any smooth $\mathbf{C}$-valued functions on $\mathfrak{H}_m$ then

$$\Delta(fg) = \sum_{r=0}^{m} \mathrm{tr}[{}^t\rho_r(\partial/\partial z)f \cdot \rho_{m-r}^*(\partial/\partial z)g] \tag{4.27}$$

(b)

$$\begin{aligned} \rho_r(\partial/\partial z)\det(z)^\alpha &= c_r(\alpha)\det(z)^{\alpha-1}\rho_{m-r}^*(\partial/\partial z), \\ \rho_r^*(\partial/\partial z)\det(z)^\alpha &= c_r(\alpha)\det(z)^{\alpha-1}\rho_{m-r}(\partial/\partial z) \end{aligned} \tag{4.28}$$

for $\alpha \in \mathbf{C}$ with

$$c_r(\alpha) = \prod_{k=0}^{r-1}(\alpha + (k/2)) = \frac{\Gamma_r(\alpha + (r+1)/2)}{\Gamma_r(\alpha + (r-1)/2)} \tag{4.29}$$

Now we define polynomial functions $\lambda_r(X)$ of $X \in \mathrm{M}_m(\mathbf{C})$ by

$$\det(z1_m - X) = \sum_{r=0}^{m}(-1)^r \lambda_r(X) t_{m-r},$$

with a variable t. Then

$$\begin{aligned} &\lambda_r(X) = \mathrm{tr}[\rho_r(X)], \\ &\Delta(e^{\mathrm{tr}(uz)}\det(z)^\alpha) = e^{\mathrm{tr}(uz)}\det(z)^{\alpha-1}\sum_{r=0}^{m} c_{m-r}(\alpha)\lambda_r(uz). \end{aligned} \tag{4.30}$$

To prove (4.30) it is sufficient to use (4.27) with $f = e^{\mathrm{tr}(uz)}, g = \det(z)^\alpha$ and the apply formulas (4.28). Indeed,

$$\begin{aligned} &\Delta(e^{\mathrm{tr}(uz)}\det(z)^\alpha) = \\ &\sum_{r=0}^{m} \mathrm{tr}[{}^t\rho_r(\partial/\partial z)e^{\mathrm{tr}(uz)} \cdot \rho^*_{m-r}(\partial/\partial z)\det(z)^\alpha] = \\ &\sum_{r=0}^{m} c_{m-r}(\alpha)\det(z)^{\alpha-1}\mathrm{tr}[\rho_r(z(\partial/\partial z)e^{\mathrm{tr}(uz)})] = \\ &\sum_{r=0}^{m} c_{m-r}(\alpha)\det(z)^{\alpha-1}\lambda_r(z(\partial/\partial z)e^{\mathrm{tr}(uz)})] = \\ &\sum_{r=0}^{m} c_{m-r}(\alpha)\det(z)^{\alpha-1}\lambda_r(zu) = \sum_{r=0}^{m} c_{m-r}(\alpha)\lambda_r(uz)\det(z)^{\alpha-1} \end{aligned}$$

(we note that $\mathrm{tr}({}^tab) = \mathrm{tr}({}^tba)$ and that for each $u = {}^tu \in \mathrm{GL}_m(\mathbf{C})$ one has $u = {}^tu \in \mathrm{GL}_m(\mathbf{C})$).

In a very similar way we get the following formula for the action of the operator, which is obtained by iteration of Δ:

$$\begin{aligned} &\Delta^n[e^{\mathrm{tr}(uz)}\det(z)^\alpha] = e^{\mathrm{tr}(uz)}\det(z)^{\alpha-n} \times \\ &\times \sum_{r_1,\cdots,r_n=0}^{m} c_{m-r_1}(\alpha)c_{m-r_2}(\alpha-1)\cdots c_{m-r_n}(\alpha-n+1)\lambda_{r_1}(uz)\cdots\lambda_{r_n}(uz). \end{aligned} \tag{4.31}$$

From this formula we deduce now the following very explicit expression for the polynomial $P(v, u; n, \beta)$, which provides us with all desired informaton about it.

4.8. Theorem. *Under the assumptions and notation as in 4.5 and 4.6 the following equality holds:*

$$P(v.u;n,\beta) = \sum_{r_1,\cdots,r_n=0}^{m} C(\beta;r_1,\cdots,r_n)\mathrm{tr}[\rho_{r_1}(v)\rho^*_{m-r_1}(u)]\cdots\mathrm{tr}[\rho_{r_n}(v)\rho^*_{m-r_n}(u)], \tag{4.32}$$

where

$$C(\beta;r_1,\cdots,r_n) = c_{m-r_1}(-\beta)c_{m-r_2}(-\beta-1)\cdots c_{m-r_n}(-\beta-n+1)\times$$
$$\times(-1)^{mn+r_1+\cdots+r_n}\frac{c_{r_1}(\kappa-k)\cdots c_{r_1}(\kappa-k-n+1)}{c_m(\kappa-k)\cdots c_m(\kappa-k-n+1)},$$

and for $r_1 = r_2 = \ldots = r_n = m$ *we have* $C(\beta;m,\cdots,m) = 1$;

$$\kappa = (m+1)/2, \quad c_r(\alpha) = \prod_{k=0}^{r-1}(\alpha + (k/2)) = \frac{\Gamma_r(\alpha+(r+1)/2)}{\Gamma_r(\alpha+(r-1)/2)} = (-1)^r\frac{\Gamma_r(1-\alpha)}{\Gamma_r(-\alpha)}.$$

From this formula it easy to see that $P(v,u;n,\beta) \in \mathbf{Q}[u,v]$ *for* $\beta \in \mathbf{Q}$ *and that*

$$P(v,u;n,\beta) \equiv \mathrm{det}v^n(\mathrm{mod}\langle u_{ij}\rangle$$

(in view of properties of matrix coefficients of the representations ρ_r, ρ^*_r *which are homogenuous polynomials of the degree* m*).*

The proof of the theorem 4.8 is then an immediate consequence of the equalities (4.26). Namely, in view of (4.31) we have

$$R(z;n,\beta) = \sum_{r_1,\cdots,r_n=0}^{m}(-1)^{r_1+\cdots+r_n}c_{m-r_1}(-\beta)c_{m-r_2}(-\beta-1)\cdots c_{m-r_n}(-\beta-n+1)\times$$
$$\times\lambda_{r_1}(z)\lambda_{r_2}(z)\cdots\lambda_{r_n}(z),$$

and

$$\frac{\Gamma_m(k-(m+1)/2)}{\Gamma_m(k-n-(m+1)/2)} = (-1)^{mn}c_m(\kappa-k)\cdots c_m(\kappa-k-n+1),$$

and according to formulas (4.28)

$$\lambda_r(v\cdot\partial/\partial u)\mathrm{det}u^\alpha = \mathrm{tr}[\rho_r(v\cdot\partial/\partial u)\mathrm{det}u^\alpha] = c_r(\alpha)\mathrm{tr}[\rho_r(v)\mathrm{det}(u)^{\alpha-1}\rho^*_{m-r}(u)].$$

This completes the proof of both theorem 4.8 and 4.6.

Chapter 3: Non-Archimedean standard zeta functions of Siegel modular forms

In this chapter we give explicit formulas for the special values of the standard zeta function $\mathcal{D}(s,f,\chi)$ of a Siegel cusp form f of even degree m and of weight $k > 2m+2$ with a Dirichlet character ψ; using these formulae we construct then a non-Archimedean interpolation of the special values. For convenience, we introduce normalized standard zeta functions

$$\mathcal{D}^*(s,f,\chi),\quad \mathcal{D}^+(s,f,\chi),\quad \mathcal{D}^-(s,f,\chi),$$

multiplying $\mathcal{D}(s,f,\chi)$ by certain Γ-factors; then we formulate our results in terms of these functions. First we state a theorem on holomorphy properties of the function $\mathcal{D}^*(s,f,\chi)$ (theorem 1.3). This theorem provides a generalization of a result of A.N.Andrianov and V.L.Kalinin [An-K] on analytic properties of the standard zeta function; it is proved in §3 by use of a detailed study of poles and residues of Siegel – Eisenstein series done by Shimura [Shi10] and P.Feit [Fe]. Then we turn to algebraic properties of the special values

$$\frac{\mathcal{D}^+(s,f,\chi)}{\langle f,f\rangle}\quad \text{for } s = 1,2,\cdots,k-\nu-m$$

and

$$\frac{\mathcal{D}^-(s,f,\chi)}{\langle f,f\rangle}\quad \text{for } s = 1-k+\nu+m,\cdots,-1,0,$$

where $\langle f,f\rangle$ denotes the Petersson scalar product, $\nu = 0,1$ according as $\chi(-1) = (-1)^\nu$ (Theorem 1.4). The main result of the chapter is stated as Theorem 1.6 establishes non-Archimedean interpolation of these special values by means of the theory of non-Archimedean integration. We construct the non-Archimedean standard zeta functions $\mathcal{D}^{c+}(x,f)$, $\mathcal{D}^{c-}(x,f)$ as the S-adic Mellin transform of certain measures obtained from the special values. In their turn, these measures come from complex valued distributions $D_s^\pm(\chi) = D_s^\pm(\chi,f)$ of §2. After regularization given in §4 these distributions become bounded measures taking (p-adic) algebraic values at certain points. We then obtain the non-Archimedean zeta functions in Theorem 1.6 as the non-Archimedean Mellin transforms of these regularized measures.

§1.Description of the non-Archimedean standard zeta functions

1.1. The domain of definition of our S-adic zeta functions is the p-adic analytic Lie group

$$X_S = \mathrm{Hom}_{\mathrm{contin}}(\mathbf{Z}_S^\times, \mathbf{C}_p^\times)$$

of all p-adic characters of the group $\mathbf{Z}_S^\times$ where $\mathbf{C}_p = \widehat{\overline{\mathbf{Q}}}_p$ is the Tate field (completion of an algebraic closure of the p-adic field $\mathbf{Q}_p$), so that all integers k can be regarded as the characters $x_p^k : y \mapsto y^k$. Put

$$U = \{x \in \mathbf{Z}_p^\times | x \equiv 1 (\mathrm{mod} p^\nu)\},$$

where $\nu = 1$ or 2 according as $p > 2$ or $p = 2$, then there is the decomposition

$$X_S = X((\mathbf{Z}/p^\nu\mathbf{Z})^\times \times \prod_{q \neq p} \mathbf{Z}_q^\times) \times X(U). \tag{1.1}$$

The analytic structure on X_S is defined by means of the isomorphism

$$\phi : X(U) \xrightarrow{\sim} T = \{t \in \mathbf{C}_p^\times \,|\, |t-1|_p < 1\},$$

with $\phi(x) = x(1+p^\nu), 1+p^\nu$ being topological generator of the multiplicative group U (see §4 of chapter 1). Elements χ of the torsion subgroup $X_S^{\text{tors}} \subset X_S$ form a discrete subgroup and can be regarded as primitive Dirichlet characters χ with $S(\chi) \subset S$, where $S(\chi)$ is the support $S(C(\chi))$ of the conductor of χ (see 4.2 of chapter 1). Recall that every bounded C_p-analytic function F over X_S is uniquely determined by its values $F(\chi_0\chi)$ with χ_0 fixed and χ being taken in X_S^{tors} with possible exclusion of a finite number of them in each analyticity component of the decomposition (1.1). This condition is satisfied, for example, by the set of characters $\chi \in X_S^{\text{tors}}$ with a S-complete conductor (i.e. with $S(\chi) = S$) and even under the additional assumption that χ^2 is non trivial (this remark is essentially used in the proof of the uniqueness property of the theorem 1.6).

Let μ be a bounded $\mathbf{C}_p$-valued measure on $\mathbf{Z}_p^\times$, then its non-Archimedean Mellin transform is given by

$$L_\mu(x) = \mu(x) = \int_{\mathbf{Z}_S^\times} x \, d\mu(x) \quad (x \in X_S), \tag{1.2}$$

and determines a bounded $\mathbf{C}_p$-analytic function

$$L_\mu : X_S \to \mathbf{C}_p$$

(see §3 of chapter 1, and [Man4], [Man6], [V1])

1.2. Let

$$f = \sum_{\xi > 0} a(\xi) e_m(\xi z)$$

be the Fourier expansion of a Siegel cusp form of the even degree m, the weight k, on the congruence subgroup

$$\Gamma_0^m(C) = \left\{ \begin{pmatrix} a & b \\ c & d \end{pmatrix} \in \mathrm{Sp}_m(\mathbf{Z}) | c \equiv 0 \bmod C \right\}$$

with a Dirichlet character ψ mod C. Suppose that f is an eigenfunction of the global Hecke algebra

$$\mathcal{L}^m(C) = \otimes_{q \nmid C} \mathcal{L}_q^m(C) \tag{1.3}$$

with the eigenvalue given by a homomorphism $\Lambda : \mathcal{L}^m(C) \to \mathbf{C}$ (i.e. $f|X = \Lambda(X)f$ for all $X \in \mathcal{L}_q^m(C)$). Let

$$\alpha_0(q), \alpha_1(q), \cdots, \alpha_m(q) \tag{1.4}$$

be a $(m+1)$-tuple of Satake q-parameters, which uniquely determine Λ so that the relation

$$\alpha_0^2(q)\alpha_1(q)\cdots\alpha_m(q) = \psi(q)^m q^{m(k-(m+1)/2)} \tag{1.5}$$

holds. Recall that the *standard zeta function* of f with a Dirichlet character χ mod N is defined as the following Euler product

$$\mathcal{D}(s,f,\chi) = \prod_{q, q \nmid C} \mathcal{D}^{(q)}(s,f,\chi), \tag{1.6}$$

with

$$D^{(q)}(s,f,\chi)^{-1} =$$
$$\left(1 - \frac{\chi(q)\psi(q)}{q^s}\right) \prod_{i=1}^{m} \left(1 - \frac{\chi(q)\psi(q)\alpha_i(q)}{q^s}\right)\left(1 - \frac{\chi(q)\psi(q)\alpha_i(q)^{-1}}{q^s}\right),$$

where the product converges absolutely for $\mathrm{Re}(s) > 1+m$. Also, let us consider, together with (1.5), the following three types of the normalized zeta functions

$$\mathcal{D}^*(s,f,\chi) = (2\pi)^{-m(s+k-(m+1)/2)}\Gamma((s+\delta)/2)\prod_{j=1}^{m}\Gamma(s+k-j)\mathcal{D}(s,f,\chi), \tag{1.7}$$

$$\mathcal{D}^-(s,f,\chi) = (2\pi)^{-m(s+k-(m+1)/2)}\prod_{j=1}^{m}\Gamma(s+k-j)\mathcal{D}(s,f,\chi), \tag{1.8}$$

$$\mathcal{D}^+(s,f,\chi) = \frac{2i^\delta\Gamma(s)\cos(\pi(s-\delta)/2)}{(2\pi i)^s}\mathcal{D}^-(s,f,\chi) \tag{1.9}$$

where $\delta = 0$ or 1 according as $\psi\chi(-1) = 1$ or -1.

1.3. Theorem *(on analytic properties of the standard zeta functions). Let χ be a Dirichlet character modulo a positive integer N (not necessarily primitive) and*

$$f = \sum_{A_m \ni \xi > 0} a(\xi)e_m(\xi z) \in \mathcal{S}_k^m(C,\psi)$$

be a Siegel cusp form of weight $k \geq m+\nu$ where $\nu = 0,1$ with $\chi(-1) = (-1)^\nu$, and assume that the condition $C\det 2\xi_0$ divides N is satisfied for some matrix ξ_0 such that $a(\xi_0) \neq 0$. Then the function $\mathcal{D}^*(s,f,\chi)$ admits an analytic continuation which is holomorphic for all $s \in \mathbf{C}$ with the possible exclusion of a simple pole at the point $s = 1$ in case when the character $\chi^2\psi^2$ is trivial.

This theorem is proved in §3 by means of the detailed study of poles and residues of the Siegel – Eisenstein series as functions of the variable s (3.7 of chapter 2) using the techniques of [Fe], [Shi10].

1.4. Theorem *(Algebraic properties of the special values of the standard zeta functions)*

a). For all integers s with $1 \le s \le k - \nu - m$ and Dirichlet series χ such that χ^2 is non-trivial for $s = 1$ we have that

$$\langle f,\ f\rangle^{-1} G(\psi\chi)^{m-1}\,\mathcal{D}^{+}(s,f,\chi) \in K = \mathbf{Q}(f,\Lambda_f,\psi,\chi),$$

where $K = \mathbf{Q}(f,\Lambda_f,\psi,\chi)$ denotes the field generated by Fourier coefficients of f, by the eigenvalues $\Lambda_f(X)$ of Hecke operators X on f, and by the values of the characters χ and ψ.

b). For all integers s with $1 - k + \nu + m \le s \le 0$ we have that

$$\langle f,\ f\rangle^{-1} G(\psi\chi)^{m}\,\mathcal{D}^{-}(s,f,\chi) \in K = \mathbf{Q}(f,\Lambda_f,\psi,\chi).$$

Remark. *(a). In view of the definitions (1.7) – (1.9) and theorem 1.3 it is easy to notice that the assumptions of theorem 1.4 imply vanishing of the following special values:*

$$\mathcal{D}^{+}(s,f,\chi) = 0 \quad \text{for} \quad s \in \mathbf{N}, \quad s \not\equiv \delta(\mathrm{mod}2)$$

and

$$\mathcal{D}^{-}(s,f,\chi) = 0 \quad \text{for} \quad s \equiv \delta(\mathrm{mod}2), \quad s \in \mathbf{Z}, \quad s \le 0.$$

(b). From the proof of theorem 1.4 in §3 (see 3.15) one can substract a more explicit information about the action of Galois automorphisms $\sigma \in \mathrm{Aut}(\mathbf{C})$ on the special values (1.10), (1.11). Namely, for some non-zero constant $\mu(\Lambda,k,\psi) \in \mathbf{C}^{\times}$ depending only on $k > 2m+2$, the character ψ and the homomorphism $\Lambda : \mathcal{L}^m(C) \to \mathbf{C}$ from theorem 1.4 we have that

$$\left[\frac{G(\psi\chi)^{m-1}\mathcal{D}^{+}(s,f,\chi)}{\mu(\Lambda,k,\psi)}\right]^{\sigma} = \frac{G(\psi^{\sigma}\chi^{\sigma})^{m-1}\mathcal{D}^{+}(s,f^{\sigma},\chi^{\sigma})}{\mu(\Lambda^{\sigma},k,\psi^{\sigma})} \tag{1.12}$$

$$\left[\frac{G(\psi\chi)^{m}\mathcal{D}^{-}(s,f,\chi)}{\mu(\Lambda,k,\psi)}\right]^{\sigma} = \frac{G(\psi^{\sigma}\chi^{\sigma})^{m}\mathcal{D}^{-}(s,f^{\sigma},\chi^{\sigma})}{\mu(\Lambda^{\sigma},k,\psi^{\sigma})} \qquad (1.13).$$

Also, the following equality holds

$$\mu(\Lambda,k,\psi)^{-1}\langle f,f\rangle_C \ \in \mathbf{Q}(f,\Lambda,\psi),$$

(this means that the Petersson scalar product $\langle f,f\rangle_C$ differs from the constant $\mu(\Lambda,k,\psi)$ only by an algebraic multiple from the field $K_0 = \mathbf{Q}(f,\Lambda,\psi)$ generated by Fourier coefficients of f and the values of Λ,ψ). In (1.12), (1.13) we adopted the standard notation

$$f^{\sigma} = \sum_{A_m \ni \xi > 0} a(\xi)^{\sigma} e_m(\xi z) \in S_k^m(C,\psi^{\sigma})$$

for the action of $\sigma \in \mathrm{Aut}(\mathbf{C})$.

The theorem 1.4 is proved in §3 (theorem 3.2), and the algebraicity properties, analogous to (1.12), (1.13) are established in 3.15. For the special values s in theorem 1.4 analogous properties where studies in ealier works of S.Böcherer [Bö2], [Bö4], M.Harris [Har1] and J.Sturm [St2].

1.5. Before giving the precise statement of the main result of the chapter we make some additional assumptions on f. First of all we assume that f is p-ordinary (with respect to a fixed embedding $i_p : \overline{\mathbf{Q}} \hookrightarrow \mathbf{C}_p$). This means that

$$|i_p(\alpha_0(q)|_p = 1 \quad \text{for all} \quad q \in S. \tag{1.14}$$

Of course, for $q \neq p$ this condition is automaticaly satisfied because of the relation (1.14), and for $q = p$ we may take in (1.14) any of the numbers

$$\alpha_0(p)\alpha_{i_1}(p)\cdots\alpha_{i_r}(p) \quad (1 \le i_1 < \cdots < i_r \le m),$$

instead of $\alpha_0(q)$ since these numbers are permutated by the action of Weyl group W_m described in 1.6 of chapter 2 (the last property is easily verified by applying several substitutions of the type

$$\alpha_0(p) \mapsto \alpha_0(p)\alpha_i(p), \quad \alpha_i(p) \mapsto \alpha_i(p)^{-1}, \alpha_j(p) \mapsto \alpha_j(p) \quad (j \neq i, \; i = 1, 2, \cdots, m))$$

Recall that these numbers are inverse roots of the characteristic p-polynomial for the *spinor zeta function* $Z(s, f)$ of the cusp form f (see §1 of chapter 2):

$$\begin{aligned} &Z^{(p)}(s,f)^{-1} = Q_{f,p}(p^{-s}) = \\ &1 - \lambda_f(p)p^{-s} + \cdots + \psi(p)^m p^{2^{m-1}(mk-m(m+1)/2)} p^{-2^m s} = \\ &(1-\alpha_0(p)p^{-s}) \prod_{r=1}^{m} \prod_{1\le i_1<\cdots<i_r\le m} (1-\alpha_0(p)\alpha_{i_1}(p)\cdots\alpha_{i_r}(p)p^{-s}), \end{aligned}$$

where

$$\lambda_f(p) = \lambda_f(T(p)), \quad T((p)) = \sum_{\nu(g)=p} (\Gamma g \Gamma)$$

is the Hecke operator for the group $\Gamma = \Gamma_0^m(C), {}^t g J_m g = \nu(g) J_m$.

For each $q \not| C$ we fix any $(m+1)$-tuple satisfying (1.14) and extend the functions $\alpha_i(n)$ to all $n \in \mathbf{N}$, $(n, C) = 1$ by multiplicativity. We may suppose without loss of generality that $a(\xi) \in \overline{\mathbf{Q}}$ for all $\xi > 0$; this condition is always satisfied by elements of a certain basis in the subspace of Hecke eigenforms in $S_k^m(N, \psi)$ with the given eigenvalue $\Lambda = \Lambda_f$ (which is a character of the Hecke algebra). Indeed, this property is deduced from formulas for the action of the Hecke algebra on Fourier expansions (see (1.40), (1.41)) which is defined over $\overline{\mathbf{Q}}$. Let us assume also that

$$a(\xi_0) = 1 \tag{1.15a}$$

for some $\xi_0 > 0$ with $\det 2\xi_0 = 1$. Moreover, we make the essential assumption that

$$a_0(\xi_0) \neq 0 \tag{1.15b}$$

for the Fourier coefficient $a_0(\xi_0)$ of the S-auxiliary form

$$f_0(z) = \sum_{\xi > 0} a_0(\xi) e_m(\xi z)$$

defined in §1 of chapter 2 (see (1.47)). According to its definition, this form satisfies the following multiplicativity property

$$a_0(M\xi) = \alpha_0(M)a_0(\xi) \text{ for } M \in \mathbf{N} \text{ with } S(M) \subset S.$$

The importance of the above non-vanishing condition (1.15b) of $a_0(\xi_0)$ was pointed out to the author by S. Böcherer. It turns out that it actually may not be satisfied for certain Eisenstein series.

Suppose also that the fixed set S of prime numbers contains the support (i.e. all prime divisors) of the number $2\det(2\xi_0)$ (q_0 denotes the level of ξ_0), and our last assumption is that

$$S \cap S(C) = \emptyset, \text{ i. e. } (C, M_0) = 1, \text{ with } M_0 = \prod_{q \in S} q. \tag{1.16}$$

Set

$$N_0 = 4q_0 M_0^{\tilde{m}-1} C, \quad \tilde{m} = 2^m.$$

1.6. Main theorem. *Let*

$$f(z) = \sum_{\xi > 0} a(\xi) e_m(\xi z) \in S_k^m(C, \psi)$$

be a cusp form of weight k with Dirichlet character ψ mod C which is an eigenfunction for the Hecke algebra $\mathcal{L}^m(C)$ with eigenvalues given by a homomorphism $\Lambda : \mathcal{L}^m(C) \to \mathbf{C}$. Suppose that for the cusp form f the conditions (1.14) – (1.16) are satisfied. Then for a positive integer $c > 1$ with the condition $(c, N_0) = 1$ there exist bounded analytic functions

$$\mathcal{D}^{c+}(x, f), \mathcal{D}^{c-}(x, f) : \ X_S \to \mathbf{C}_p, \tag{1.17}$$

which are uniquely determined by the following conditions:

a) for all Dirichlet characters $\chi \in X_S^{\mathrm{tors}}$ with an S-complete conductor C_χ (i.e. with $S(\chi) = S$) and for all integers s with $1 \le s \le k - \nu - m$ the following equality holds

$$\begin{aligned} &\mathcal{D}^{c+}(\chi x_p^s, f) = \\ &i_p\Big[\frac{G_m(1_m, \chi)\, C_\chi^{m(s+k-1-m)}}{\alpha_0(C_\chi)^2} \frac{C(\overline{\chi}\psi)^s}{G(\overline{\chi}\psi)}(1 - (\overline{\chi}\psi)^2(c)c^{-2s}) \times \\ &\times \prod_{q|C} \{(1 - (\chi\overline{\psi})_0(q)q^{s-1})/(1 - (\overline{\chi}\psi)_0(q)q^{-s})\} \frac{\mathcal{D}^+(s, f, \overline{\chi})}{\langle f,\ f\rangle}\Big], \end{aligned} \tag{1.18}$$

b) for all Dirichlet characters $\chi \in X_S^{\mathrm{tors}}$ with $S(\chi) = S$ and for all integers s with $1 - k + \nu + m \le s \le 0$ the following equality holds

$$\begin{aligned} &\mathcal{D}^{c-}(\chi x_p^s, f) = \\ &i_p\Big[\frac{G_m(1_m, \chi)\, C_\chi^{m(s+k-1-m)}}{\alpha_0(C_\chi)^2}(1 - (\chi\overline{\psi})^2(c)c^{2s-2}) \frac{\mathcal{D}^-(s, f, \overline{\chi})}{\langle f,\ f\rangle}\Big], \end{aligned} \tag{1.19}$$

where

$$G_m(\xi,\chi) = \sum_{h\in \mathbf{M}_m(\mathbf{Z}) \bmod C_\chi} \chi(\det h) e_m(\xi h/C_\chi)$$

denotes the Gauss sum of degree m of the primitive Dirichlet character $(\psi\overline{\chi})_0$ mod $C_{\psi\overline{\chi}}$ *associated with* $\psi\overline{\chi}$ mod $C\,C_\chi$ *with the normalized zeta functions* $\mathcal{D}^{\pm}(s,f,\chi)$ *being defined by (1.8), (1.9), and for a Dirichlet character* χ $\nu = 0$ *or 1 according as* $\chi(-1) = (-1)^\nu$, ξ_0 *being a fixed half integral symmetric positive definite matrix satisfying the above assumptions.*

1.7. *Proof* of the main theorem and of theorems 1.3 and 1.4 is based on the relation of the function $\mathcal{D}(s,f,\chi)$ with a convolution of Rankin type given by the general Andrianov's identity:

$$\begin{aligned}
&2a(\xi_0)\det\xi_0^{-(s+k-1+\nu)/2}\mathcal{D}(s,f,\chi) = L_{C\,C_\chi}(s+\frac{m}{2},\psi\chi_{\xi_0}\chi)\times \\
&\times \prod_{i=0}^{(m/2)-1} L_{C\,C_\chi}(2s+2i,\psi^2\chi^2))L((s+k-1+\nu)/2,f,\theta^{(\nu)}_{2\xi_0}(\chi)),
\end{aligned} \tag{1.20}$$

where χ is a Dirichlet character, $\nu = 0$ or 1 according as $\chi(-1) = (-1)^\nu$, ξ_0 being a fixed half integral symmetric positive definite matrix. In this identity the left hand side of (1.20) is an Euler product associated with the homomorphism $\Lambda : \mathcal{L}^m(C) \to \mathbf{C}$, and the right hand side of (1.20) is completely determined by the Fourier coefficients of f. The non-Archimedean part of the construction is based on the theory of distributions and S-adic integration.

Using a general criterion of finite additivity we construct complex valued distributions associated with $\mathcal{D}(s,f,\chi)$ by defining their values at Dirichlet characters in §2. We prove in §3 an (Archimedean) integral representation for these values, which enables us to express the distributions in terms of the Fourier coefficients of Siegel – Eisenstein series from §3 of the previous chapter by applying the holomorphic projection operator (see §4 of chapter 2). After a regularization given in §4 these distributions become bounded $\mathbf{C}_p$-valued measures $\mathcal{D}^{c+}$, $\mathcal{D}^{c-}$ taking algebraic values at compact open subsets of $\mathbf{Z}_S$. The proof of the main theorem is then completed by application of the non-Archimedean Mellin transform. The uniqueness of the non-archimedean standard zeta functions is a consequence of the remark in the end of 1.1.

§2. Complex valued distributions associated with standard zeta functions of Siegel modular forms

2.1. Let as in 1.2

$$f = \sum_{A_m \ni \xi > 0} a(\xi) e_m(\xi z) \in \mathcal{S}_k^m(C,\psi) \tag{2.1}$$

be a cusp form of degree m weight k with the Dirichlet character ψ modulo C on the congruence subgroup $\Gamma_0^m(C)$, which is an eigenfunction of all local Hecke algebras $\mathcal{L} = \mathcal{L}_q^m$ with q not dividing the level C of f, so that the Satake q-parameters $\alpha_i = \alpha_i(q)$ $(i = 0,1,\cdots,m)$ are defined for all $q \nmid C$. We extend by multiplicativity the definition of $\alpha_i(M)$ to all $M \in \mathbf{N}$, M prime to the level C.

Now for the fixed set of prime numbers $S = \{q\}$ not dividing C we define complex valued distributions on the profinite group

$$G_S = \prod_{q \in S} \mathbf{Z}_q^{\times} \tag{2.2}$$

associated with the functions $\mathcal{D}(s, f, \chi)$. The crucial role in the construction is played by the cusp form (1.50), chapter 2:

$$f_0 = f_{0,S} = \sum_{M | M_0^{\tilde{m}-1}} \alpha_0(M)^{-1} f | V^+(M). \tag{2.3}$$

with $M_0 = \prod_{q \in S} q$, $\tilde{m} = 2^m$. Recall that if

$$f_{0,S}(z) = \sum_{A_m \ni \xi > 0} a_0(\xi) e_m(\xi z) \in \mathcal{M}_m^k(CM_0^{\tilde{m}-1}, \psi) \tag{2.4}$$

be Fourier expansion of the function $f_{0,S}(z)$ then the coefficients $a_0(\xi)$ satisfy the following multiplicativity property: for all $M \in \mathbf{N}$ with $S(M) \subset S$

$$a_0(M\xi) = \alpha_0(M) a_0(\xi) \quad (\xi \in A_m, \ \xi \geq 0). \tag{2.5}$$

Our construction of the distributions is based on the general Andrianov's identity (1.20) expressing $\mathcal{D}(s, f, \chi)$ in terms of the Rankin type convolution:

$$\begin{aligned} 2a(\xi_0)\det\xi_0^{-(s+k-1+\nu)/2}\mathcal{D}(s, f, \chi) &= L_{CM}(s + \frac{m}{2}, \psi\chi_{\xi_0}\chi) \times \\ \times \prod_{i=0}^{(m/2)-1} L_{CM}(2s + 2i, \psi^2\chi^2)) &L((s + k - 1 + \nu)/2, f, \theta_{2\xi_0}^{(\nu)}(\chi)), \end{aligned} \tag{2.6}$$

where $\chi \bmod M$ is a Dirichlet character modulo M, $\nu = 0, 1$, $\chi(-1) = (-1)^\nu$, $A_m \ni \xi_0$ is some appropriate half integral symmetric positive definite matrix, $s \in \mathbf{C}$ is a complex number with $\mathrm{Re}(s) \gg 0$,q_0 is the level of ξ_0, $q_0 \in \mathrm{M}_m(\mathbf{Z})$. Put

$$\hat{\xi}_0 = q_0 \xi_o^{-1}, N_0 = 4q_0 C M_0^{\tilde{m}-1}.$$

2.2. Proposition. *Let $s \in \mathbf{C}$, $\mathrm{Re}(s) \gg 0$. Then there exists a complex valued distribution $\mathcal{D}_s$ on G_S which is uniquely determined by its values on Dirichlet characters $\chi \bmod M$ with $S(M) \subset S$ given by*

$$\begin{aligned} &2a_0(\xi_0)\det\xi_0^{-(s+k-1+\nu)/2}\mathcal{D}_s(\chi) = \\ &\alpha_0(M_0^{\tilde{m}-1}M')^{-1}(CM_0^{\tilde{m}-1}M')^{m(2s+2k-2-m)/4}C^{m(2\nu+m)/4} \times \\ &L_{N_0}(s + \frac{m}{2}, \psi\chi_{\xi_0}\overline{\chi}) \prod_{i=0}^{(m/2)-1} L_{N_0}(2s + 2i, \psi^2\overline{\chi}^2) \times \\ &\times \det((2q_0)^{-1/2}\xi_0)^{(m/2)+\nu} L((s + k - 1 + \nu)/2, f_0 | V(C), \theta_{2\hat{\xi}_0}^{(\nu)}(\chi_M) | W(N_0 M')), \end{aligned} \tag{2.7}$$

where

$$f_0|V(C)(z) = f_0(Cz),$$
$$\theta^{(\nu)}_{2\hat{\xi}_0}(\chi_M)|W(N_0M')(z) = \det(\sqrt{N_0M'z})^{-(m/2)-\nu}\theta^{(\nu)}_{2\hat{\xi}_0}(\chi_M)(-(N_0M'z)^{-1}), \tag{2.8}$$

where M, M' are sufficiently large positive integers with the condition

$$M_0C_\chi|M, \quad MM_0C_\chi^2|M'$$

so that $S(M) = S(M') = S$ with C_χ being the conductor of the character χ and χ_M the Dirichlet character modulo M induced by χ.

Proof. According to the general criterion of finite additivity applied for the system of functions $\mathcal{D}_{s,M} : (\mathbf{Z}/M\mathbf{Z})^\times \to \mathbf{C}$ (see [Ka3], [Maz-SD] and (3.4) of chapter 1) it suffices to verify that the right hand side of (2.7) does not depend on M and M'. Here it is essential to take into account the non-vanishing condition (1.15b). The independence of M obviously follows from the S-completeness of M (i.e. M is divisible by all primes from S). In order to show that this expression also does not depend on M' we put

$$M' = AB, \ A = M_0^2C_\chi^2, \ B = M_1 = M'(M_0C_\chi)^{-2}$$

and use the equality

$$\theta^{(\nu)}_{2\hat{\xi}_0}(\chi_M)|W(N_0M') = M_1^{m(2\nu+m)/4}g|V(M_1), \tag{2.9}$$

in which

$$g(z) = \theta^{(\nu)}_{2\hat{\xi}_0}(\chi_M)|W(N_0M_0^2C_\chi^2)(z) = \sum_{A_m\ni\xi\geq 0} b(\xi)e_m(\xi z).$$

Now from definition (2.23) of chapter 2 and from (2.9) it follows that

$$L(s, f_0|V(C), \theta^{(\nu)}_{2\hat{\xi}_0}(\chi_M)|W(N_0M')) = M_1^{m(2\nu+m)/4}L(s, f_0|V(C), g|V(M_1)) =$$
$$M_1^{m((2\nu+m)/4-s)}C^{-ms}\sum_{\substack{A_m\ni h>0\\ \mathrm{mod}\sim}}\varepsilon(\xi)^{-1}a_0(M_1\xi)b(C\xi)\det\xi^{-s}. \tag{2.10}$$

The fact, that (2.7) does not depend on M' (or, equvalently, on M_1) follows now from the multiplicativity property (2.5): $a_0(M\xi) = \alpha_0(M)a_0(\xi)$. If we substitute (2.10) into (2.7) with s equal to $(s+k-1+\nu)/2$, $M' = M_0^2C_\chi^2M_1$ we see that M_1 cancels, proving the proposition .

Another way of proving the proposition is to calculate explicitly the integrals $\mathcal{D}_{s,M}(\chi)$ of the Dirichlet characters χ which is done below under some additional technical assumptions.

2.3. Proposition. *Let $\chi \bmod M$ be a primitive Dirichlet character with an S-complete conductor C_χ, $C_\chi|M$ (i.e. $S(\chi) = S(C_\chi) = S$). Assume also that $S(2\det\xi_0) \subset S$. Then the following identity holds*

$$\mathcal{D}_{s,M}(\chi) = \frac{G_{C_\chi}(1_m,\chi)}{C_\chi^{m^2/2}}\cdot\frac{C_\chi^{m(2s+2k-2-m)/4}}{\alpha_0(C_\chi)^2}\mathcal{D}(s,f,\overline{\chi}) \tag{2.11}$$

where

$$G_{C_\chi}(1_m,\chi) = \sum_{h\in M_m(\mathbb{Z}) \bmod C_\chi} \chi(\det h) e_m({}^t\xi h/C_\chi)$$

is the Gauss sum of degree m of the character χ.

The proof of the proposition 2.3 is based on the transformation formula (2.4) (of chapter 2) for the theta function

$$\theta_{\xi_0}^{(\nu)}(\chi)|W(q_0Q^2) = \chi(-1)^m \frac{G_{C_\chi}(1_m,\chi)}{C_\chi^{m^2}} \det(q_0^{-1/2}\xi_0)^{(m/2)+\nu}\theta_{\hat{\xi}_0}^{(\nu)}(\overline{\chi}), \tag{2.12}$$

If we now take into account that

$$G_{C_\chi}(1_m,\chi)G_{C_\chi}(1_m,\overline{\chi}) = \chi(-1)^m C_\chi^{m^2}, \quad \theta_{2\xi_0}^{(\nu)}(\chi) = \theta_{\xi_0}^{(\nu)}(\chi)|V(2),$$

then (2.12) transforms to

$$\theta_{2\hat{\xi}_0}^{(\nu)}(\chi)|W(4q_0Q^2) = \chi(-1)^m \frac{G_{C_\chi}(1_m,\chi)}{C_\chi^{m^2}} \det(q_0^{1/2}\xi_0^{-1})^{(m/2)+\nu}\theta_{2\xi_0}^{(\nu)}(\overline{\chi}), \tag{2.13}$$

Next we write:

$$N_0M' = 4q_0Q^2N' \ , \ N' = M_0^{\tilde{m}-1}CM'C_\chi^{-2},$$

so that

$$\theta^{(\nu)}|W(N_0M') = (N')^{m(2\nu+m)/4}\left(\theta^{(\nu)}|W(4q_0Q^2)\right)|V(N') \tag{2.14}$$

and if we substitute (2.14) into (2.7) we get exactly the desired identity, proving the proposition.

Now we establish an integral representation for the values of the distributions $\mathcal{D}_{s,M}$ with $\mathrm{Re}(s) \gg 0$ using the identity (2.29) of chapter 2, which implies that for all Dirichlet characters χ mod M with the condition $M_0|M$ the following hold:

$$\begin{aligned} &2a(\xi_0)((4\pi)^m\det(\xi_0))^{-(s+k-1+\nu)/2}\Gamma_m((s+k-1+\nu)/2)\,\mathcal{D}(s,f,\chi) = \\ &\quad L_{N_0}(s+\frac{m}{2},\psi\chi_{\xi_0}\chi)\prod_{i=0}^{(m/2)-1} L_{N_0}(2s+2i,\psi^2\chi^2))\times \\ &\quad \times\langle f^\rho(z),\theta_{2\xi_0}^{(\nu)}(z;\chi)E(z,(s-k+m+\nu)/2)\rangle_N, \end{aligned} \tag{2.15}$$

where

$$E(z,s) = E(z,s;k-(m/2)-\nu,\psi\chi_{\xi_0}\chi,N)$$

is the Siegel – Eisenstein series of the weight $k-(m/2)-\nu$, the level $N = 4q_0CM^2$, and the Dirichlet character $\psi\chi_{\xi_0}\chi$ (see 2.4 of chapter 2).

2.4. Proposition. *For $s \in \mathbf{C}$, $\mathrm{Re}(s) \gg 0$ there is the following integral representation for the distributions of proposition 2.2:*

$$2a_0(\xi_0)((4\pi)^m \det\xi_0)^{-(s+k-1+\nu)/2}\Gamma_m((s+k-1+\nu)/2)\mathcal{D}_{s,M}(\chi) =$$
$$C^{m(s+k-1+\nu)/2}\alpha_0(M_0^{\tilde{m}-1}M')^{-1}(M_0^{\tilde{m}-1}M')^{m(2s+2k-2-m)/4}\det(q_0^{-1/2}\xi_0)^{(2\nu+m)/2}\times$$
$$\times L_{N_0}(s+\frac{m}{2},\psi\chi_{\xi_0}\overline{\chi}) \prod_{i=0}^{(m/2)-1} L_{N_0}(2s+2i,\psi^2\overline{\chi}^2))\times$$
$$\times \langle f_0^{\rho}(z)|V(C),\theta_{2\hat{\xi}_0}^{(\nu)}(\chi_M)|W(N_0M')E(z,(s-k+m+\nu)/2)\rangle_N, \tag{2.16}$$

where M, M' are sufficiently large positive integers satysfying the conditions

$$M_0C_\chi|M, \quad MM_0C_\chi^2|M'$$

(so that $S(M) = S(M') = S$), C_χ is the conductor of χ, $C_\chi|M$, χ_M the Dirichlet character modulo M induced by χ, and

$$E(z,s) = E(z,s;k-(m/2)-\nu,\psi\chi_{\xi_0}\overline{\chi},N) \tag{2.17}$$

is the Siegel –Eisenstein series of the weight $k-(m/2)-\nu$, the level $N = N_0CM'$, and the Dirichlet character $\psi\chi_{\xi_0}\overline{\chi}$ (see 2.4).

The proof of the proposition follows from the the general identity (2.15)in which we take f to be equal to the cusp form $f_0|V(C) \in S_k^m(C^2M_0^{\tilde{m}-1},\psi)$ of level $C^2M_0^{\tilde{m}-1}$ defined by (2.7).

The next important ingredient of our construction is an application of the trace operator to modular forms of level $N = CN_0M'$ in the above integral formula, which enables us to reduce all considerations (e.g. the integral formula) to the case of modular forms of the fixed level CN_0.

2.5. The trace operator. We define the trace operator $\mathrm{Tr}_{CN_0}^{CN_0M'}$ acting on (C^∞-)modular forms of the degree m, weight k, and the Dirichlet character $\overline{\psi}$ on the congruence subgroup $\Gamma_0^m(CN_0M')$, by the following equality:

$$F|\mathrm{Tr}_{CN_0}^{CN_0M'} = \sum_{{}^tu=u\in \mathrm{M}_m(\mathbf{Z}) \bmod M'} F|\begin{pmatrix} 1_m & 0 \\ N_0Cu & 1_m\end{pmatrix}. \tag{2.18}$$

Then the scalar product in (2.16) transforms in the obvious way:

$$\langle f_0^\rho, F\rangle_{CN_0M'} = \langle f_0^\rho, F|\mathrm{Tr}_{CN_0}^{CN_0M'}\rangle_{CN_0}. \tag{2.19}$$

Now we give a useful description of action of the trace operator on Fourier expansions in terms of the operator $U(M')$ whose action on Fourier expansions is given by the formula:

$$F|U(M')(z) = (M')^{-m(m+1)/2} \sum_{{}^tu=u\in \mathrm{M}_m(\mathbf{Z}) \bmod M'} F((z+u)/M'). \tag{2.20}$$

(in the notation of the chapter 2 this operator coincides with the Frobenius operator $\Pi_+(M')$ extended by multiplicativity to positive integral values of M' prime to the level C of the form F: $F|U(M') = F|\Pi_+(M'))$. If

$$F(z) = \sum_{\xi \in A_m} a(\xi, y) e_m(\xi z),$$

then

$$F|U(M')(z) = \sum_{\xi \in A_m} a(M'\xi, {M'}^{-1} y) e_m(\xi z). \tag{2.21}$$

If we take into account the matrix identity

$$\begin{pmatrix} 1_m & 0 \\ CN_0 u & 1_m \end{pmatrix} = (CN_0 M')^{-1} \begin{pmatrix} 0 & -1_m \\ CN_0 M' 1_m & 0 \end{pmatrix} \begin{pmatrix} 1_m & -u \\ 0 & M' 1_m \end{pmatrix} \begin{pmatrix} 0 & -1_m \\ CN_0 1_m & 0 \end{pmatrix},$$

we see that the trace operator satisfy the following relation:

$$F|\mathrm{Tr}_{CN_0}^{CN_0 M'} = (M')^{-m(k-m-1)/2} F|W(CN_0 M') U(M') W(CN_0). \tag{2.22}$$

Now let us apply (2.19) and (2.22) to the integral representation (2.16) then we get

$$2a_0(\xi_0)((4\pi)^m \det \xi_0)^{-(s+k-1+\nu)/2} \Gamma_m((s+k-1+\nu)/2)\, \mathcal{D}_{s,M}(\chi) =$$

$$\alpha_0 (M_0^{\tilde{m}-1} M')^{-1} (M_0^{\tilde{m}-1} M')^{m(2s+2k-2-m)/4} \det(q_0^{-1/2} \xi_0)^{(2\nu+m)/2} \times \tag{2.23}$$

$$\times L_{N_0}(s + \frac{m}{2}, \psi \chi_{\xi_0} \overline{\chi}) \prod_{i=0}^{(m/2)-1} L_{N_0}(2s + 2i, \psi^2 \overline{\chi}^2)) \times$$

$$\times (M')^{m(m+1-k)/2} \langle f_0^{\rho}(z)|V(C), F(s,\chi)|U(M')W(CN_0)\rangle_{CN_0},$$

where

$$F(s,\chi) = \theta_{2\hat{\xi}_0}^{(\nu)}(\chi_M)|V(C) \cdot E(z, (s-k+m+\nu)/2)|W(CN_0 M'),$$

and we took into account the following transformation formula:

$$\theta_{2\hat{\xi}_0}^{(\nu)}(\chi_M)|W(N_0 M')|W(CN_0 M') = (-1)^{m(m/2+\nu)} C^{-m(2\nu+m)/4} \theta_{2\hat{\xi}_0}^{(\nu)}(\chi_M)|V(C). \tag{2.24}$$

§3. Algebraic properties of the special values of normalized distributions

3.1. In this section we consider only the case of even m. In order to give a precise statements on algebraic properties of the standard zeta functions and of the corresponding distributions, it is convenient to make some additional normalization of these values because these properties look different for the integral points to the left and to the right of the critical line $\mathrm{Re}(s) = \frac{1}{2}$ (the same effect happens in the case of the Riemann zeta function).

Recall that as in §1 we have introduced the following three types of the normalized zeta function: for $\kappa = (m+1)/2$, $s \in \mathbf{C}$

$$\mathcal{D}^*(s,f,\chi) = \\ (2\pi)^{-m(s+k-\kappa)}\Gamma((s+\delta)/2)\prod_{j=1}^{m}\Gamma(s+k-j)\mathcal{D}(s,f,\chi), \tag{3.1}$$

$$\mathcal{D}^-(s,f,\chi) = \Gamma((s+\delta)/2)^{-1}\mathcal{D}^*(s,f,\chi) = \\ (2\pi)^{-m(s+k-\kappa)}\prod_{j=1}^{m}\Gamma(s+k-j)\mathcal{D}(s,f,\chi), \tag{3.2}$$

$$\mathcal{D}^+(s,f,\chi) = \frac{2\,i^\delta\Gamma(s)\cos(\pi(s-\delta)/2)}{(2\pi i)^s}\mathcal{D}^-(s,f,\chi) = \\ \frac{i^\delta\pi^{(1-2s)/2}}{\Gamma((1+\delta-s)/2)}\mathcal{D}^*(s,f,\chi), \tag{3.3}$$

where $\delta = 0$ or 1 according as $\psi\chi(-1) = (-1)^\delta$ with $f \in \mathcal{S}_k^m(C,\psi)$ be an eigenfunction of the global Hecke algebra $\mathcal{L}^m(C) = \otimes_{q\nmid C}\mathcal{L}_q^m(C)$ with the eigenvalue given as a homomorphism $\Lambda : \Lambda^m(C) \to \mathbf{C}$. The convenience of the $*$-normalization (3.1) is explained by the fact that the standard zeta function (continued holomorphically onto the entire complex plane) satisfies a certain functional equation relating

$$\pi^{-s/2}\mathcal{D}^*(s,f,\chi) \quad \text{and} \quad \pi^{-(1-s)/2}\mathcal{D}^*(s,f^\rho,\overline{\chi}). \tag{3.4}$$

However, the precise form of this functional equation is known only in few cases [An-K], [Bö4]. The principial difficulty in dealing with the general case is that for $m > 1$ one lacks the correct and reasonable definition of the Euler factors for the standard zeta function at bad primes, i.e. for $q|C$. In the one dimentional case these factors are determined by the Atkin – Lehner theory [At-Le], [At-Li], [La3], [Li1]; however, the similar theory does not exist yet even in the case $m = 2$.

Now we turn to the normalizations (3.3) and (3.3). Their convenience is illustrated by the following result on algebraicity of the special values of (3.2) and (3.3) which is proven in this section together with the corresponding statement about the special values of normalized distributions given below in 3.3.

3.2. Theorem *(Algebraic properties of the special values of standard zeta functions). Assume that the cusp eigenform $f \in \mathcal{S}_k^m(C,\psi)$ satisfies the condition $a(\xi_0) = 1$ for some $\xi_0 \in A_m, \xi_0 > 0$. Then*

a) For all integers s satisfying $1 \le s \le k-\delta-m$ and $s \ne 1$ if the character $\chi^2\psi^2$ is trivial we have that

$$\langle f,\ f\rangle^{-1}\,\mathcal{D}^+(s,f,\chi) \in K = \mathbf{Q}(f,\Lambda_f,\psi,\chi), \tag{3.5}$$

where $K = \mathbf{Q}(f,\Lambda_f,\psi,\chi)$ denote the field generated by Fourier coefficients of f, by the eigenvalues $\Lambda_f(X)$ of Hecke operators X on f, and by the values of the characters χ and ψ. Moreover,

$$\mathcal{D}^+(s,f,\chi) = 0 \quad \text{for} \quad s \not\equiv \delta (\mathrm{mod} 2).$$

b) For all integers s satisfying $1-k+\delta+m \leq s \leq 0$ we have that

$$\langle f,\ f\rangle^{-1}\mathcal{D}^{-}(s,f,\chi) \in K, \tag{3.6}$$

and

$$\mathcal{D}^{-}(s,f,\chi)=0 \quad \text{for} \quad s \equiv \delta(\text{mod} 2), \quad s \in \mathbf{Z}, s \leq 0.$$

The proof of the theorem is completed in 3.15, although we assume in the proof that χ is a (not necessarily primitive) Dirichlet character modulo N such that $2C\det 2\xi_0$ divides N for a fixed matix $\xi_0 \in A_m, \xi_0 > 0$ satisfying the condition $a(\xi_0) \neq 0$. This assumption is insignificant and can be dropped if we multiply the special values in question by a finite number of Euler factors corresponding to the divisors of $\det 2\xi_0$; these factors take algebraic values in K for arbitrary integers s.

3.3. Now we define the normalized distributions by the following formula

$$\langle f,f\rangle_C \mathcal{D}^{*}_{s,M}(\chi) = (2\pi)^{-m(s+k-(m+1)/2)}\Gamma((s+\delta)/2)\prod_{j=1}^{m}\Gamma(s+k-j)\mathcal{D}_{s,M}(\chi), \tag{3.7}$$

$$\mathcal{D}^{-}_{s,M}(\chi) = \langle f,f\rangle_C^{-1}(2\pi)^{-m(s+k-(m+1)/2)}\prod_{j=1}^{m}\Gamma(s+k-j)\mathcal{D}_{s,M}(\chi), \tag{3.8}$$

$$\mathcal{D}^{+}_{s,M}(\chi) = \frac{2i^{\delta}\Gamma(s)\cos(\pi(s-\delta)/2)}{(2\pi i)^{s}}\mathcal{D}^{-}_{s,M}(\chi), \tag{3.9}$$

where $\mathcal{D}_{s,M}(\chi)$ are the values of the distribution $\mathcal{D}_{s,M}$ on the profinite group

$$G_S = \mathbf{Z}_S^{\times} = \prod_{q\in S}\mathbf{Z}_q^{\times} \xrightarrow{\sim} \mathrm{Gal}(\mathbf{Q}(S)/\mathbf{Q})$$

with $\mathrm{Gal}(\mathbf{Q}(S)/\mathbf{Q})$ being the Galois group of the maximal abelian extension of $\mathbf{Q}$ unramified outside S and ∞. In the definitions (3.7) - (3.9) we use notation χ_M for the Dirichlet character modulo M induced by χ and $S(M) \subset S$.

The following proposition is closely related to theorem 2.2.

3.4. Proposition. *(Algebraicity properties of values of the normalized distributions) Assume that the cusp eigenform $f \in S_k^m(C,\psi)$ satisfies the conditions $a(\xi_0)=1$ for some $\xi_0 \in A_m, \xi_0 > 0$ and $a(\xi) \in \overline{\mathbf{Q}}$ for all $\xi > 0$. Then*
a) For all integers s such that $1 \leq s \leq k-\nu-m$ and $s \neq 1$ if the character $\chi^2\psi^2$ is trivial we have that

$$\mathcal{D}^{+}_{s,M}(\chi) \in K = \mathbf{Q}(f,\Lambda_f,\psi,\chi,\alpha_i(q); i \leq m, q \in S),$$

where $K=\mathbf{Q}(f,\Lambda_f,\psi,\chi)$ denotes the field generated by Fourier coefficients of f, by the eigenvalues $\Lambda_f(X)$ of Hecke operators X on f, the Satake q-parameters and by the values of the characters χ and ψ.
b) For all integers s such that $1-k+\nu+m \leq s \leq 0$ we have that

$$\mathcal{D}^{-}_{s,M}(\chi) \in K.$$

In the proof of (a) and (b) we assume that the set S contains all of the prime divisors of $2\det(2\xi_0)$. According to the explicit formula of proposition 2.3 for Dirichlet characters χ with S-complete conductors the values $\mathcal{D}^{\pm}_{s,M}(\chi)$ are expressed in terms of the corresponding special values of the normalized standard zeta functions $\mathcal{D}^{\pm}(s, f^{\rho}, \overline{\chi})$ so that in this case the proposition is an immediate consequence of theorem 3.2. However, for the non-Archimedean construction we have to consider all the characters of finite order of G_S because are involved in the generalized Kummer congruences in §4. Very explicit formulas for the corresponding values of the distributions are given in 3.6, these formulas can not be simply reduced to those in theorem 3.2. The proof of both the algebraicity properties and the proof of theorem 3.2 is completed in 3.15 . Application of a fixed embedding $i_p : \overline{\mathbf{Q}} \hookrightarrow \mathbf{C}_p$ to the normalized distributions $\mathcal{D}^{\pm}_{s,M}(\chi)$ gives us certain p-adic distributions $i_p[\mathcal{D}^{\pm}_{s,M}(\chi)]$ which become bounded p-adic measures after a regularization in §4. The basic fact used in the proof of proposition 3.4 is an integral representation for $\mathcal{D}^{\pm}_{s,M}(\chi)$ which we state now in a preliminary form in terms of the distribution $\mathcal{D}^{*}_{s,M}(\chi)$.

3.5. Proposition. *Let $f \in S^m_k(C,\psi)$, m be even, χ be a Dirichlet character modulo $M \geq 1$, and assume that the non-vanishing condition (1.15b) for the Fourier coefficient $a_0(\xi_0)$ is satisfied. Then there is the following integral representation for the normalized distributions:*

$$\langle f, f\rangle_C \mathcal{D}^{*}_{s,M}(\chi) = \gamma(M')\langle f_0^{\rho}| V(C), F^{*}_{M'}(s,\chi)| W(CN_0)\rangle_{CN_0}, \tag{3.10}$$

where

$$\gamma(M') = 2^{m(2k-2-m-\kappa)} i^{-m(k-(m/2)-\nu)} a_0(\xi_0)^{-1} \alpha_0(M_0)^{\tilde{m}-1} M'^{-1} (C M_0^{\tilde{m}-1})^{(k-1-m)/2},$$

$$F^{*}_{M'}(s,\chi) = (q_0 C)^{-m(2s+m)/4} \det\xi_0^{(s+k-1+\nu)/2} \det(q_0^{-1/2}\xi_0)^{(2\nu+m)/2} \times$$
$$\times [2^{-1}\theta^{(\nu)}_{2\hat{\xi}_0}(\chi_M)| V(C)\, G^{*}(z, s-k+\nu+m)/2)]| U(M'),$$

and

$$G^{*}(z,s) = G^{*}(z,s;k-(m/2)-\nu, \overline{\chi}\chi_{\xi_0}\psi, N)$$

is the normalised Siegel – Eisenstein series defined in §3 of the chapter 2 with $N = CN_0M' = C^2 M_0^{\tilde{m}-1} 4 q_0 M'$, and the Dirichlet character $\overline{\chi}\chi_{\xi_0}\psi$ mod N.

The right hand side of (3.10) is well defined and holomorphic for all $s \in \mathbf{C}$ with the possible exclusion of $s = 1$ in the case when the character $\overline{\chi}^2\psi^2$ is trivial.

The proof of the proposition follows from the definition of the normalized Siegel – Eisenstein series $G^{*}(z,s)$ which for k equal to $k-(m/2)-\nu$, s equal to $(s-k+\nu+m)/2$

takes the following form

$$
\begin{aligned}
&G^*(z,(s-k+m+\nu)/2) = \\
&(4q_0C^2M')^{m(2s+m)/4}\tilde{\Gamma}(k-(m/2)-\nu,(s-k+\nu+m)/2)\times \\
&\times L_N(s+(m/2),\overline{\chi}\chi_{\xi_0}\psi)\prod_{j=1}^{m/2} L_N(2s+m-2j,\psi^2\overline{\chi}^2)\times \\
&\times 2^{m\kappa}E(z,(s-k+\nu+m)/2)|W(N),
\end{aligned}
\tag{3.11}
$$

where

$$
\begin{aligned}
&\tilde{\Gamma}(k-(m/2)-\nu,(s-k+\nu+m)/2) = \\
&i^{m(k-(m/2)-\nu)}2^{-m(k-(m/2)-\nu+1)}\pi^{-m(s+k-\nu)/2}\Gamma_m((s+k-\nu)/2)\Gamma((s+\delta)/2),
\end{aligned}
$$

and it suffices to take into account the following relation for the gamma factors:

$$
\begin{aligned}
&(2\pi)^{-m(s+k-\kappa)}\Gamma((s+\delta)/2)\prod_{j=1}^{m}\Gamma(s+k-j) = \\
&2^{m(2k-2-m-\kappa)}i^{-m(k-(m/2)-\nu)}\tilde{\Gamma}(k-\nu-(m/2),(s-k+\nu+m)/2)\times \qquad (3.12). \\
&\times 2^{m(m+1)/2}2^{m(2s+m)/2}(4\pi)^{-m(s+k-1+\nu)/2}\Gamma_m((s+k-\nu-m)/2),
\end{aligned}
$$

The relation (3.12) follows from the duplication formula

$$
\Gamma(\frac{s}{2})\Gamma(\frac{s+1}{2}) = 2^{1-s}\sqrt{\pi}\Gamma(s),
$$

and the definition

$$
\Gamma_m(s) = \pi^{(m-1)m/4}\prod_{j=0}^{m-1}\Gamma(s-(j/2)).
$$

The statement on holomorphy is implied by the theorem on holomorphy of the Siegel – Eisenstein series $G^*(z,s)$ (see theorem 3.6 of the chapter 2) and by the uniform estimates on $z \in \mathfrak{H}_m$ for Fourier coeficients, from which it follows that the scalar product in the right hand side of (3.10) can be defined for all $s \in \mathbf{C}$ $s \neq 1$, if $\psi^2\overline{\chi}^2$ is trivial.

Now we specialize the integral representation (3.10) to the case of the critical values s (see 3.3 of chapter 2), when the confluent hypergeometric function admits an elementary expression in terms of a certain polynomial (see Theorem 4.8 of chapter 2). Then we use the holomorphic projection operator $\mathcal{H}ol$ (see §4 of chapter 2 in order to obtain an integral representation for the distributions $\mathcal{D}^{\pm}_{s,M}(\chi)$ in terms of the holomorphic Siegel modular forms with algebraic (and explicitly given) Fourier coefficients.

3.6. Proposition. *Under the assumptions and notations of Proposition 3.4 the following integral representations are valid* *(a)For all integers s with $1 \leq s \leq k-\delta-m$ such that for $s=1$ the character $\chi^2\overline{\psi}^2$ is non-trivial we have that*

$$\begin{aligned} &\langle f,f\rangle_C\mathcal{D}^+_{s,M}(\chi)=\gamma(M')\langle f_0^\rho|V(C),F^+_{M'}(s,\chi)|W(CN_0)\rangle_{CN_0}, \\ &F^+_{M'}(s,\chi)=\sum_{A_m\ni>0}\ \sum_{C\hat{\xi}_0[h_1]+h_2=M'h} d^+(s,h_1,h_2)e_m(hz); \end{aligned} \tag{3.13}$$

(b) For all integers s with $1-k+\delta+m \leq s \leq 0$ we have that

$$\begin{aligned} &\langle f,f\rangle_C\mathcal{D}^-_{s,M}(\chi)=\gamma(M')\langle f_0^\rho|V(C),F^-_{M'}(s,\chi)|W(CN_0)\rangle_{CN_0}, \\ &F^-_{M'}(s,\chi)=\sum_{A_m\ni>0}\ \sum_{C\hat{\xi}_0[h_1]+h_2=M'h} d^-(s,h_1,h_2)e_m(hz); \end{aligned} \tag{3.14}$$

The functions $F^\pm_{M'}(s,\chi)\in\mathcal{M}(CN_0,\psi)$ are holomorphic Siegel modular forms with cyclotomic Fourier coefficients explicitly given by:
if $s\not\equiv\delta(\mathrm{mod}2)$, $1\leq s\leq k-\nu-m$ then $d^+(s,h_1,h_2)=0$;
if $s\equiv\delta(\mathrm{mod}2)$, $1\leq s\leq k-\nu-m$ then

$$\begin{aligned} d^+(s,h_1,h_2)&=\chi_M(\det h_1)\det h_1^\nu \det h_2^{(2s-1)/2}P(h_2,h,s)\times \\ &\times L_N^+(s,\overline{\chi}\chi_{\xi_0}\psi\xi_{h_2})M(h,\overline{\chi}\chi_{\xi_0}\psi,s+(m/2))\times \\ &\times(q_0C)^{-m(2s+m)/4}\det\xi_0^{(s+k-1+\nu)/2}\det(q_0^{-1/2}\xi_0)^{(2\nu+m)/2}; \end{aligned} \tag{3.15}$$

if $s\equiv\delta(\mathrm{mod}\,2)$, $1-k+\nu+m\leq s\leq 0$ then $d^-(s,h_1,h_2)=0$;
if $s\not\equiv\delta(\mathrm{mod}\,2)$, $1-k+\nu+m\leq s\leq 0$ then

$$\begin{aligned} d^-(s,h_1,h_2)&=\chi_M(\det h_1)\det h_1^\nu P(h_2,h,1-s)\times \\ &\times L_N^-(s,\overline{\chi}\chi_{\xi_0}\psi\xi_{h_2})M(h,\overline{\chi}\chi_{\xi_0}\psi,s+(m/2))\times \\ &\times(q_0C)^{-m(2s+m)/4}\det\xi_0^{(s+k-1+\nu)/2}\det(q_0^{-1/2}\xi_0)^{(2\nu+m)/2}; \end{aligned} \tag{3.16}$$

$P(v,u;s)\in\mathbf{Q}[v,u]$ denotes a polynomial of entries of matrix variables $v=(v_{ij}),u=(u_{ij})$ which is defined for $s\equiv\delta(\mathrm{mod}2), 1\leq s\leq k-m-\nu$ with rational coefficients independent of f, M, χ, M' and satisfies the property

$$P(x,y,s)\equiv\det x^{(k-\nu-m-s)/2} \bmod (\langle y_{ij}\rangle)$$

$M(h,\overline{\chi}\chi_{\xi_0}\psi,s+(m/2)$ denotes the integer multiple which comes up in the Fourier expansion of the Siegel - Eisenstein series (see (3.44) of the chapter 2),

and

$$L^+_N(s,\omega) = \frac{2i^\delta\Gamma(s)\cos(\pi(s-\delta)/2)}{(2\pi)^s}L_N(s,\omega),$$

$$L^-_N(s,\omega) = L_N(s,\omega) \quad (\delta = 0,1, \ \omega(-1) = (-1)^\delta)$$

are the normalized Dirichlet L-series with the Euler factors at q, $q|N$ *removed from their Euler products. The summation in the inner sums of (3.13) and (3.14) is taken over all pairs* (h_1, h_2) *of integral matrices with the conditions*

$$h_1 \in \mathrm{M}^+_m(\mathbf{Z}), \quad h_2 > 0, \quad h_2 \in A_m, \quad C\hat{\xi}_0[h_1] + h_2 = M'h$$

(i.e. h_1 *is an integral matrix with a positive determinant (not necessarily symmetric),* $h_2 \in A_m$ *is a positive definite half integral matrix, and* $C\hat{\xi}_0[h_1]$ *denotes the matrix given by* $C^t h_1(q_0\xi^{-1})h_1 = Cq_0\xi_0^{-1}[h_1]$*).*

3.7. *The proof* of the proposition is carried out in several steps. First we write down a preliminary integral representation using proposition 2.4 and definitions (3.8), (3.9) of the normalized distributions:

$$\langle f,f\rangle_C \mathcal{D}^\pm_{s,M}(\chi) = \gamma(M')\langle f_0^\rho|V(C), \tilde{F}^\pm_{M'}(s,\chi)|W(CN_0)\rangle_{CN_0}, \tag{3.17}$$

where

$$F^\pm_{M'}(s,\chi) = (q_0C)^{-m(s+(m/2))}\det\xi_0^{(s+k-1+\nu)/2}\det(q_0^{-1/2}\xi_0)^{(2\nu+m)/2}\times$$

$$\times\, [2^{-1}\theta^{(\nu)}_{2\hat{\xi}_0}(\chi_M)|V(C)\cdot G^\pm_{M'}(z,(s-k+m+\nu)/2)]|U(M').$$

We already know the Fourier expansions of the functions in (3.17): by the definition of theta functions from §2 of chapter 2 we have that

$$\theta^{(\nu)}_{2\hat{\xi}_0}(\chi_M)|V(C) = \sum_{h_1\in \mathrm{M}_m(\mathbf{Z})+} \chi_M(\det h_1)\det h_1^\nu e_m(C\hat{\xi}_0[h_1]z). \tag{3.18}$$

The Fourier expansions of the series $G^\pm_{M'}(z,s)$ is explicitly written in §3 of chapter 2. If we substitute in the expansions (3.53), (3.53a) of chapter 2 s equal to $(s-k+\nu+m)/2$, k equal to $k-(m/2)-\nu$, we get the equality:

$$G^\pm_{M'}(z,(s-k+m+\nu)/2) = \sum_{h\in A_m} b^\pm(h,y,(s-k+\nu+m)/2)e_m(hz), \tag{3.19}$$

in which

$$b^\pm(h,y,(s-k+\nu+m)/2) =$$

$$L^\pm(s,\overline{\chi}\chi_{\xi_0}\chi_h\psi)W^*(h,y,(s-k+\nu+m)/2)M(h,\overline{\chi}\chi_{\xi_0}\psi, s+(m/2)).$$

For the critical values of s the function $W^*(h,y,(s-k+\nu+m)/2)$ is explicitly given by formulas (3.31), (3.34) of chapter 2 in terms of the polynomial

$$R(z;n,\beta) = (-1)^{mn}e^{\mathrm{tr}(z)}\det(z)^{n+\beta}\Delta^n_m[e^{\mathrm{tr}(z)}\det(z)^{-\beta}], \tag{3.20}$$

where

$$\Delta_m = \det(\partial_{ij}), \quad \partial_{ij} = 2^{-1}(1+\delta_{ij})\partial/\partial_{ij},$$

where $n \geq 0$ is an integer and β a complex number. The degree of $R(z;n,\beta)$ is equal to mn, and the term of the highest degree coincides with $\det z^n$. If $\beta \in \mathbf{Q}$ then $R(z;n,\beta) \in \mathbf{Q}[z_{ij}]$. We set

$$Q(y,s) = R(y;(k-\nu-m-s)/2,(1-s-k+\nu+m)/2) \tag{3.21}$$

then for $s \in \mathbf{Z}$, $1-k+\nu+m \leq s \leq k-\nu-m$, the coefficients of (3.19) transform to the following:
if $\leq s \leq k-\nu-m$, $s \not\equiv \delta(\bmod 2)$and $\overline{\chi}^2\psi^2$ is non trivial for $s=1$ then $b^+(h,y,(s-k+\nu+m)/2)=0$;
if $\leq s \leq k-\nu-m$, $s \equiv \delta\,(\bmod 2)$and $\overline{\chi}^2\psi^2$ is non trivial for $s=1$ then

$$b^+(h,y,(s-k+\nu+m)/2) = (\det h)^{s-\frac{1}{2}}\det(4\pi y)^{(s-k+\nu+m)/2}Q(4\pi hy,s)\times$$
$$\times L^+_M(s,\overline{\chi}\chi_{\xi_0}\chi_h\psi)M(h,\overline{\chi}\chi_{\xi_0}\psi,s+(m/2)); \tag{3.22}$$

if $s \equiv \delta(\bmod 2)$, $1 \leq s \leq k-\nu-m$ and $\psi^2\overline{\chi}^2$ is non trivial for $s=1$ then $b^-(h,y,(s-k+\nu+m)/2)=0$;
if $s \not\equiv \delta(\bmod 2)$, $1-k-s+\nu+m \leq s \leq 0$ then

$$b^-(h,y,(s-k+\nu+m)/2) = \det(4\pi y)^{(1-s-k+\nu+m)/2}Q(4\pi hy,1-s)\times$$
$$\times L^-_M(s,\overline{\chi}\chi_{\xi_0}\chi_h\psi)M(h,\overline{\chi}\chi_{\xi_0}\psi,s+(m/2)); \tag{3.23}$$

In (3.22), (3.23) we suppose that $h > 0$, $h \in A_m$. According to the theorem on positivity of matrix indices in Fourier expansions (theorem 3.7 of chapter 2) we have for these values of s that $b^+(h,y,(s-k+\nu+m)/2)=0$ if $h \in A_m$ is not positive definite. Respectively, we have for the corresponging values of s that $b^-(h,y,(s-k+\nu+m)/2)=0$ if $h \in A_m$ is non-negative (i.e. if it does not contain negative eigenvalues). However we now will see that the Fourier expansions of the functions $\tilde{F}^{\pm}_{M'}(s,\chi)$ from (3.17) involve only those Fourier coefficients of $G^{\pm}_{M'}(z,(s-k+m+\nu)/2)$ in the expansion (3.19) which correspond to terms with positive definite $h \in A_m$.

3.8. Proposition. *(a) If $s \in \mathbf{Z}$, $1 \le s \le k-\nu-m$ and $\overline{\chi}^2\psi^2$ is non trivial for $s=1$ then we have in (3.17) that*

$$\tilde{F}^+_{M'}(s,\chi) = \sum_{A_m \ni >0} \sum_{C\hat{\xi}_0[h_1]+h_2=M'h} \tilde{d}^+(s,h_1,h_2)e_m(hz); \tag{3.24}$$

(b) If $1-k+\nu+m \le s \le 0$, $s \in \mathbf{Z}$ then

$$\tilde{F}^-_{M'}(s,\chi) = \sum_{A_m \ni >0} \sum_{C\hat{\xi}_0[h_1]+h_2=M'h} \tilde{d}^-(s,h_1,h_2)e_m(hz); \tag{3.25}$$

the functions $\tilde{F}^+_{M'}(s,\chi)$, $\tilde{F}^-_{M'}(s,\chi) \in \tilde{\mathcal{M}}^m_k(CN_0,\overline{\psi})$ are non holomorphic Siegel modular forms with the Fourier coefficients given by:

$$\tilde{d}^+(y,s,h_1,h_2)e_m(hz) = \chi_M(\det h_1)\det h_1^\nu \det h_2^{(2s-1)/2}\det(4\pi y)^{(s-k+\nu+m)/2}\times$$

$$\times\, Q(4\pi hy,s)L^+_N(s,\overline{\chi}\chi_{\xi_0}\psi\xi_{h_2})M(h,\overline{\chi}\chi_{\xi_0}\psi,s+(m/2))\times \tag{3.26}$$

$$\times\,(q_0C)^{-m(2s+m)/4}\det\xi_0^{(s+k-1+\nu)/2}\det(q_0^{-1/2}\xi_0)^{(2\nu+m)/2};$$

$$\tilde{d}^-(y,s,h_1,h_2) = \chi_M(\det h_1)\det h_1^\nu\det(4\pi y)^{1-s-\nu-m+k)/2}\times$$

$$\times\, Q(4\pi hy,1-s)L^-_N(s,\overline{\chi}\chi_{\xi_0}\psi\xi_{h_2})M(h,\overline{\chi}\chi_{\xi_0}\psi,s+(m/2))\times \tag{3.27}$$

$$\times\,(q_0C)^{-m(2s+m)/4}\det\xi_0^{(s+k-1+\nu)/2}\det(q_0^{-1/2}\xi_0)^{(2\nu+m)/2}$$

where $Q(y,s)$ is the polynomial defined by (3.21), and the summation is extended to all pairs h_1,h_2 of matrices $h_1 \in \mathrm{M}^+_m(\mathbf{Z})$, $A_m \ni h_2 > 0$ satisfying the condition

$$Cq_0{}^th_1\xi_0^{-1}h_1 + h_2 = M'h, \quad A_m \ni h > 0.$$

The proof of the proposition is an immediate consequence of the preliminary integral representation (3.17) and the formulas (3.18), (3.19), (3.22), (3.23) for the Fourier coefficients. The final step in proving proposition 3.6 consists in deducing it from the already proven proposition 3.8 . For this purpose we apply the holomorphic projection operator $\mathcal{H}ol$ to the C^∞-modular forms $\tilde{F}^\pm_{M'}(s,\chi)$ and we then use the formulas from theorem 4.6 of chapter 2 which describe its action on Fourier expansions. In order to justify the applicability of this result (more precisely, the statement (b)) to our situation in order to calculate $\mathcal{H}ol\,(\tilde{F}^\pm_{M'}(s,\chi))$ we note that the positivity property of the Fourier expansion for the functions $G^\pm_{M'}(z,s)$ in proposition 3.8 implies the corresponding property for the functions $\tilde{F}^\pm_{M'}(s,\chi)$. On the other hand, the moderate growth condition follows from growth estimates given in [St2] (see also [Fe], [Shi10]). The meaning of these estimates is that for critical values of s (i.e. for which the corresponding Fourier expansions of $G^\pm_{M'}(z,s)$ contain only terms with positive definite matrix indices) these (non holomorphic) modular forms satisfy the same growth estimates that those valid for holomorphic modular forms (see (1.15),(1.16) of chapter 2). Hence we obtain the necessary growth estimates also for functions $\tilde{F}^{\pm}_{M'}(s,\chi)$. The moderate growth condition (4.3) of chapter 2 can be verified by applying to it the same upper estimate as in the integral formula (4.14) of chapter 2. It completes the proof of proposition 3.6.

Now we will prove theorem 1.3 on analytic properties of the special values of standard zeta functions. We start with an integral representation for the normalized zeta-function $\mathcal{D}^*(s,f,\chi)$ which is analogous to that for the distributions $\mathcal{D}^*_{s,f}$ (see (3.10) in proposition 3.5).

3.9. Proposition. *Let $f \in S^m_k(C,\psi)$ be a cusp form of weight $k \geq m+1$ where m is even, χ be a Dirichlet character modulo $M \geq 1$. Put $N = 4q_oM^2C$ where q_0 is the level of a quadratic form with the matrix $2\xi_0$ such that $a(\xi_0) \neq 0$. Then we have that*

$$a(\xi_0)\mathcal{D}^*_N(s,f,\chi) = \langle f^\rho, K^*(z,s;\xi_0,\chi)\rangle_N,$$

where

$$\begin{aligned} K^*(z,s;\xi_0,\chi) =& N^{-m(2s+m)/4} 2^{m(2k-m+2-\kappa)} i^{-m(k-(m/2)-\nu)} \times \\ & \times \det\xi_0^{(s+k-1+\nu)/2} \theta^{(\nu)}_{2\xi_0}(\chi) G^*(z,(s-k+\nu+m)/2)|W(N), \end{aligned}$$

where the subscript N in the notation $\mathcal{D}^*_N(s,f,\chi)$ indicates that all Euler factors corresponding to q, $q|N$ are removed from the Euler product, and the series

$$G^*(z,s) = G^*(z,s;k-(m/2)-\nu,\chi\chi_{\xi_0}\psi,N)$$

being defined in 3.5 , (3.45), of chapter 2.

The proof is deduced from the integral representation (2.15) rewritten in the form

$$\begin{aligned} & 2a(\xi_0)((4\pi)^m \det\xi_0)^{-(s+k-1+\nu)/2} \Gamma_m((s+k-1+\nu)/2)\,\mathcal{D}(s,f,\chi) = \\ & \langle f^\rho(z), \tilde{K}(z,s;\xi_0,\chi)\rangle_N, \end{aligned}$$

where

$$\begin{aligned} & \tilde{K}(z,s;\xi_0,\chi) = \\ & \quad L_N(s+\frac{m}{2},\psi\chi_{\xi_0}\chi) \prod_{i=0}^{(m/2)-1} L_N(2s+2i,\psi^2\chi^2))\times \\ & \quad \times \theta^{(\nu)}_{2\xi_0}(z;\chi)[E(z,(s-k+m+\nu)/2)|_{k-(m/2)-\nu}W(N)]|_{k-(m/2)-\nu}W(N). \end{aligned}$$

In the above equality we used definition (3.1) of the normalized zeta functions, the definition of the series $G^*(z,s)$ and the relation (3.12) for the Γ-factors. Now theorem 1.3 follows from the proposition 3.9 and theorem 3.6 of chapter 2 in which we take k to be equal to $k-(m/2)-\nu$.

Note that the function $\mathcal{D}^*_N(s,f,\chi)$ is obtained from the function $\mathcal{D}^*(s,f,\chi)$ if we multiply it by an elementary holomorphic factor; however we do not know in general how to obtain holomorphy properties of the function $\mathcal{D}^*(s,f,\chi)$ itself from the theorem, and this interesting question needs a further study. However, under the assumptions of theorem 1.3 we have that

$$\mathcal{D}^*(s,f,\chi) = \mathcal{D}^*_N(s,f,\chi),$$

hence the proof is completed.

3.10. In order to prove theorem 3.2 on algebraic properties we need an integral representation for the functions $\mathcal{D}^{\pm}(s,f,\chi)$ analogous to that of the proposition 3.9.

Let χ be a Dirichlet character modulo M and assume that all conditions of theorem 3.2 are satisfied. Put $N = 4q_M^2 C$, then we have the following integral representation

$$a(\xi_0)\mathcal{D}_N^{\pm}(s,f,\chi) = \langle f^\rho, K^{\pm}(z,s;\xi_0,\chi)\rangle_N, \tag{3.28}$$

with

$$K^{\pm}(z,s;\xi_0,\chi) = N^{-m(2s+m)/4} 2^{m(2k-2-m-\kappa)} i^{-m(k-(m/2)-\nu)} \times$$

$$\times \det \xi_0^{(s+k-1+\nu)/2} \mathcal{H}ol\,[\theta_{2\xi_0}^{(\nu)}(\chi)|W(N)G^{\pm}(z,(s-k+\nu+m)/2)]|W(N),$$

in which

$$K^{\pm}(z,s;\xi_0,\chi) \in \mathcal{M}_k^m(N,\overline{\psi}),$$

the series

$$G^{\pm}(z,s) = G^{\pm}(z,s;k-(m/2)-\nu,\chi\chi_{\xi_0}\psi,N)$$

are defined by (3.47), (3.48) of the previous chapter and the symbol $\mathcal{H}ol$ denotes the holomorphic projection operator from theorem 4.6 of chapter 2. The proof of (3.28) is carried out in exactly the same way as that of the proposition 3.9 if we take into account definitions (3.2) and (3.3) of the functions $\mathcal{D}^{\pm}(s,f,\chi)$, the definition of the series $G^{\pm}(z,\chi)$ and use the relation (3.12) for the Γ-factors. The possibility of applying $\mathcal{H}ol$ to the function

$$\theta_{2\xi_0}^{(\nu)}(\chi)|W(N)G^{\pm}(z,(s-k+\nu+m)/2) \tag{3.29}$$

by formulas of theorem 3.7 of chapter 2 is justified as in the end of 3.8 bearing in mind positivity properties of Fourier expansions of the series $G^{\pm}(z,s)$ in theorem 3.7 of chapter 2 and the growth estimates mentioned above. From these estimates it follows that the function (3.29) satisfy the bounded growth condition, and its Fourier expansion contains only terms with positive definite matrix indices.

Remark. *If $k > 2m+2$ then for $s = k-\nu-m$ the series defining the function*

$$G^{\pm}(z,0) = G^{*}(z,0;k-(m/2)-\nu,\chi\chi_{\xi_0}\psi,N)$$

is absolutely convergent so that this function is holomorphic, and we can omit the symbol $\mathcal{H}ol$ in the integral representation (3.28):

$$a(\xi_0)\Gamma^{+}(k-\nu-m)\mathcal{D}_N(k-\nu-m,f,\chi) = \langle f^\rho, K^{\pm}(z,k-\nu-m;\xi_0,\chi)\rangle_N, \tag{3.30}$$

with

$$\begin{aligned} K^{\pm}(z,k-\nu-m;\xi_0,\chi) = \\ N^{-m(2k-2\nu-m)/4} 2^{m(2k-2-m-\kappa)+m(2k-2\nu-m)/2} i^{-m(k-(m/2)-\nu)} \times \\ \times\, \theta_{2\xi_0}^{(\nu)}(\chi)[G^{\pm}(z,0)|W(N)], \end{aligned} \tag{3.31}$$

with

$$\Gamma^+(s) = (2\pi)^{-m(s+k-\kappa)} \frac{2i^\delta \Gamma(s)\cos(\pi(s-\delta)/2)}{(2\pi)^s} \prod_{j=1}^{\tilde{m}} \Gamma(s+k-j) \tag{3.32}$$

being the gamma-factor. With these values of k *and* s *the identity of A.N.Andrianov (proposition 2.8 of the chapter 2 and the equality (2.28)) takes the form:*

$$\begin{aligned} &a(\xi_0)\mathcal{D}^+(s,f,\chi_N) = a(\xi_0)\mathcal{D}^+(s,f,\chi) = \\ &\Gamma^+(s)L_N(s+(m/2),\chi\chi_{\xi_0}\psi) \prod_{i=0}^{(m/2)-1} L_N(2s+2i,\chi^2\psi^2)\times \\ &\times \sum_{h\in \mathrm{SL}_m(\mathbf{Z})\backslash \mathrm{M}_m^+(\mathbf{Z})} \chi_N(\det h)a(\xi_0[h])\det h^{-(s+k-1)}, \end{aligned} \tag{3.33}$$

where χ *be a Dirichlet character modulo* N *defined by* $\chi_N(d) = \chi(d)$ *for* $\det 2\xi_0|N$. *The series in (3.33) is absolutely convergent for* $\mathrm{Re}(s) > 1+m$ *due to the estimate*

$$|a(h)| = \mathcal{O}(\det h^{(k/2)})$$

for the Fourier coefficients.

3.11. Action of the group Aut(C) on scalar products of modular forms. Let us fix a positive integer N_1. Recall that the group $\mathrm{Aut}(\mathbf{C})$ acts on modular forms

$$f = \sum_{A_m \ni \xi \geq 0} a(\xi)e_m(\xi z) \in \mathcal{M}_k^m(N_1,\psi)$$

by the following rule:

$$f^\sigma = \sum_{A_m \ni \xi \geq 0} a(\xi)^\sigma e_m(\xi z) \in \mathcal{M}_k^m(N_1,\psi^\sigma) \quad (\sigma \in \mathrm{Aut}(\mathbf{C})),$$

and this action commutes with the action of the Hecke algebra, see [Shi5], [St2]. Consider the global Hecke algebra

$$\mathcal{L}(N_1) = \otimes_{q \nmid N_1} \mathcal{L}_q^m(N_1)$$

and suppose that $f \in \mathcal{M}(N_1,\psi)$ is an eigenfunction of the Hecke algebra $\mathcal{L}(N_1)$ with the eigenvalue given by a homomorphism $\Lambda : \mathcal{L}(N_1) \to \mathbf{C}$, i.e. $f|X = \Lambda(X)f$ for all $X \in \mathcal{L}_q^m(N_1)$ and all $q \nmid N_1$. Let $N_1|N$. We define a Λ-packet of modular forms as the following subspace of $\mathcal{S}_k^m(N,\psi)$:

$$H_k^m(\Lambda,N,\psi) = \{f \in \mathcal{S}_k^m(N,\psi) | f|X = \Lambda(X)f,\ X \in \mathcal{L}_q^m(N_1),\ q \nmid N\},$$

and put

$$\begin{aligned} H_k^m(\Lambda,\psi) &= \cup_{N\equiv 0 (\mathrm{mod}\, N_1)} H_k^m(\Lambda,N,\psi), \\ \mathcal{S}_k^m(\psi) &= \cup_{N\equiv 0 (\mathrm{mod}\, N_1)} \mathcal{S}_k^m(N,\psi). \end{aligned}$$

We know due to Shimura that the action of $\mathrm{Aut}(\mathbf{C})$ commutes with the action of the Hecke algebra [Shi5], therefore for each $\sigma \in \mathrm{Aut}(\mathbf{C})$

$$f \in H_k^m(\Lambda,\psi) \iff f^\sigma \in H_k^m(\Lambda^\sigma,\psi^\sigma), \tag{3.34}$$

where $\Lambda^\sigma(X) = \Lambda(X)^\sigma$. On the other hand, if we use the normality property of the Hecke operators with respect to the Petersson scalar product and commutativity of the Hecke algebra $\Lambda(N_1)$ we see (as in the classical case) that for a certain set of homomorphisms $\Lambda = \Lambda_1, \cdots, \Lambda_t$ there is the decomposition of $\mathcal{S}(N,\psi)$ into the orthogonal direct sum of the corresponding Λ-packets:

$$\mathcal{S}(N,\psi) = \bigoplus_{i=1}^{t} H_k^m(\Lambda_i, N, \psi). \qquad (3.34a)$$

the following proposition was established by J.Sturm ([St2], theorem 3). We state here this result in a form, which is more suitable for our applications.

3.12. Proposition. *Let m be even then for any integer k with $k > 2m+2$, a Dirichlet character ψ mod N, and a homomorphism $\Lambda : \mathcal{L}(N) \to \mathbf{C}$ there exists a non zero constant $\mu(\Lambda, k, \psi) \in \mathbf{C}^\times$ depending only on Λ, k, ψ such that*

$$\left[\frac{\langle f^\rho, g\rangle_N}{\mu(\Lambda,k,\psi)}\right]^\sigma = \frac{\langle f^{\sigma\rho}, g^\sigma\rangle_N}{\mu(\Lambda^\sigma, k, \psi^\sigma)} \qquad (3.35)$$

for all $f \in H_k^m(\Lambda, N, \psi), g \in \mathcal{M}_k^m(N, \overline{\psi})$, $\sigma \in \mathrm{Aut}(\mathbf{C})$.

Remark. *If we take in equality (3.35) g equal to f^ρ then proposition 3.12 implies that*

$$\langle f, f\rangle_N \mu(\Lambda, k, \psi)^{-1} \in \mathbf{Q}(\Lambda, f),$$

with $\mathbf{Q}(\Lambda, f)$ being the subfield of $\mathbf{C}$ generated by the values of the homomorphism Λ and the Fourier coefficients of f.

We give here a proof based on the Andrianov's identity (3.33) in which the right--hand side has the form:

$$\mathcal{D}^+(s, f, \chi_N) = \Gamma^+(s) \prod_{q|N} R(\Lambda, q, k)(\chi(q)\psi(q)q^{-s}) \qquad (3.36)$$

where

$$R(\Lambda, q, k)(t) \in \mathbf{Q}[\Lambda(X), X \in \mathcal{L}_q^m(N)]\,[t]$$

are polynomials satisfying $R(\Lambda, q, k)(0) = 1$ which depend only on the Λ-packet of the form f and on the numbers q and k. The product in (3.36) converges absolutely for $\mathrm{Re}(s) > 1 + m$. Put $s = k - \nu - m$ where $\nu = 0, 1$, $k \equiv \nu \bmod 2$, take as χ_N the trivial character modulo N and define

$$\mu(\Lambda, k, \psi) = G(\psi)^{m-1}\Gamma^+(k-\nu-m) \prod_{q \nmid 2B} R(\Lambda, q, k)(\psi(q)q^{-(k-\nu-m)}), \qquad (3.37)$$

with the Γ-factor being defined by (3.32), $B = B(\Lambda, k, \psi)$ being a positive integer such that the product in (3.37) does not vanish; B exists due to absolute convergence of the product (see the remark in 3.10), and we can and will assume that $B(\Lambda, k, \psi)^\sigma = B(\Lambda^\sigma, k, \psi^\sigma)$.

Now, with the number $\mu(\Lambda, k, \psi)$ already been defined, we prove first the proposition 3.12 for the special modular form $g = G(\psi)^{m-1}K^+(\xi_0, \psi)$, where we adopt the notation $K^+(\xi_0, \psi) = K^+(z, k-\nu-m; \xi_0, \chi_N)$ in order to stress, that the function

$K^+(z, k-\nu-m; \xi_0, \chi_N)$ depends on ξ_0 and on χ_N (recall that we took χ_N to be the trivial character modulo N). Then the following identity holds:

$$[G(\psi)^{m-1}K^+(\xi_0, \psi)]^\sigma = G(\psi^\sigma)^{m-1}K^+(\xi_0^\sigma, \psi^\sigma). \tag{3.38}$$

This important fact expresses in a more precise form the result of proposition 3.8 of chapter 2 on the cyclotomicity of the Fourier coefficients of this Siegel modular form. The identity (3.38) will be proved later in 3.14, and now we deduce from it proposition 3.12. According to the equalities (3.36) and (3.30), there is the following relation:

$$\frac{G(\psi)^{m-1}\langle f^\rho, K^+(\xi_0, \psi)\rangle_N}{\mu(\Lambda, k, \psi)} = a(\xi_0) \prod_{\substack{q|2B \\ q\nmid \det 2\xi_0}} R(\Lambda, q, k)(\psi(q)q^{-(k-\nu-m)}), \tag{3.39}$$

with the finite Euler product in the right hand side on which the automorphisms $\sigma \in \mathrm{Aut}(\mathbf{C})$ act term-by-term. Therefore, from (3.38) and (3.39) it follows that for the functions of the type $g = G(\psi)^{m-1}K^+(\xi_0, \psi)$ satisfy the relation (3.35).

In order to deal with the general case we vary $\xi_0 \in A_m$, ξ_0 so that the number $N = N(\xi_0)$ will now depend on ξ_0, and consider the trace operator

$$\mathrm{Tr}(N_2, N_1, \psi) = \sum_{i=1}^{d} \psi(a(i))F|_k g(i), \tag{3.40}$$

where elements $g(i) = \begin{pmatrix} a(i) & b(i) \\ c(i) & d(i) \end{pmatrix}$ form a complete system of representatives of the right cosets:

$$\Gamma_0^m(N_1) = \bigcup_{i=1}^{d} \Gamma_0^m(N_2)g(i).$$

The important property of this action is that it commutes with the trace operator. This fact is stated more precisely in the following proposition.

3.13. Proposition. *Let $F \in \mathcal{M}_k^m(N_2, \psi)$ be a Siegel modular form with cyclotomic Fourier coefficients,*

$$F(z) = \sum_{\xi \geq 0} A(\xi)e_m(\xi z), \quad A(\xi) \in \mathbf{Q}^{\mathrm{ab}}.$$

Then for all $\sigma \in \mathrm{Aut}(\mathbf{C})$ the following equality holds

$$[f|\mathrm{Tr}(N_2, N_1, \psi)]^\sigma = [f^\sigma|\mathrm{Tr}(N_2, N_1, \psi^\sigma)].$$

The proof is given in [St2], lemma 11, and it is based on properties of the action of the restricted adele group $G_{\mathbf{A}+}$ on the graded ring of automorphic forms studied by Shimura [Shi5], where $G_{\mathbf{A}+}$ denotes the subgroup of the adelization

$$G_{\mathbf{A}} = \{\alpha \in \mathrm{GL}_{2m} | {}^t\alpha J_m \alpha = \nu(\alpha)J_m,\ \nu(\alpha) \in \mathrm{GL}_1(\mathbf{A})\}$$

consisting of all elements $\alpha \in G_{\mathbf{A}}$ for which Archimedean component of the idele $\nu(\alpha)$ is positive. For $x \in G_{\mathbf{A}+}$ and a modular form $F \in \mathcal{M}_k^m = \bigcup_M \mathcal{M}_k^m(\Gamma^m(M))$ the action of x on F is denoted by F^x. If $t \in \prod_q \mathbf{Z}_q^+$ an idele whose action on $\mathbf{Q}^{ab}$ by class field theory coincides with that of $\sigma \in \mathrm{Aut}(\mathbf{C})$, then we have that

$$F^\sigma = F^{x(t^{-1})} \quad \text{with} \quad x(t) = \begin{pmatrix} 1_m & 0_m \\ 0_m & t1_m \end{pmatrix} \in G_{\mathbf{A}+}$$

for the above action on modular forms. Now the proposition is easily deduced from this equality. In view of the strong approximation theorem for the group $\mathrm{Sp}_m(\mathbf{A})$ we can choose for each representative $g(i)$ in (3.40) elements $u(i) \in \mathrm{Sp}_m(\mathbf{A})$, $h(i) \in \mathrm{Sp}_m(\mathbf{Z})$ such that $u(i)_q \equiv 1_{2m}(\mathrm{mod} N_2)$ for all primes q, $q|N_2$ and

$$\begin{pmatrix} 1_m & 0_m \\ 0_m & t^{-1}1_m \end{pmatrix} g(i) \begin{pmatrix} 1_m & 0_m \\ 0_m & t1_m \end{pmatrix} = u(i)\, h(i).$$

Let us take into account that $F^\sigma = F^{x(t^{-1})}$ then we get the equality

$$[(F^\sigma)|\mathrm{Tr}(N_2, N_1, \psi^\sigma]^{\sigma^{-1}} = [\sum_i \psi(a(i))^\sigma F^\sigma |g(i)]^{\sigma^{-1}}$$

and from the choice of $h(i)$ it follows that

$$\Gamma_0^m(N_1) = \cup_{i=1}^d \Gamma_0^m(N_2) h(i),$$

with $h(i) = \begin{pmatrix} a'(i) & b'(i) \\ c'(i) & d'(i) \end{pmatrix}$ and $a(i)' \equiv a(i) \bmod N_2$, so that the proposition follows.

Now we are able to accomplish the proof of the proposition 3.12. We let the element ξ_0 in the equality (3.39) vary, and put $N_2 = B^2 N \det^2 2\xi_0$, $N_1 = N$.Then

$$\langle f^\rho, K^+(\xi_0, \psi)\rangle_{N_2} = \langle f^\rho, K^+(\xi_0, \psi)|\mathrm{Tr}(N_2, N_1, \overline{\psi})\rangle_{N_1},$$

and from proposition 3.13 it follows that the equality (3.35) holds for all modular forms g from the set

$$\mathcal{V} = \left\{ G(\psi)^{m-1} K^+(\xi_0, \psi)|\mathrm{Tr}(N_2, N_1, \overline{\psi}) \middle| A_m \ni \xi_0 > 0, N_2 = B^2 \det^2 2\xi_0,\ N_1 = N \right\}.$$

Let

$$\mathcal{V}_1 = \left\{ g_1 \in H_k^m(\Lambda, N, \overline{\psi}) \middle| g - g_1 \text{ is orthogonal to some } g \in \mathcal{V} \right\}.$$

In other words, the set $\mathcal{V}_1$ consists of those elements in $H_k^m(\Lambda, N, \overline{\psi})$ which are orthogonal projections of the special elements $g \in \mathcal{V}$ as above. We can check as in [St2] that $\mathcal{V}_1$ generates the whole Λ-packet $H_k^m(\Lambda, N, \overline{\psi})$. Indeed, if

$$f_1 = \sum_{\xi > 0} a_1(\xi) e_m(\xi z) \in H_k^m(\Lambda, N, \overline{\psi})$$

satisfies $\langle f_1, g_1\rangle_N = 0$ for all $g_1 \in \mathcal{V}_1$ then (3.39) implies that $a(\xi_0) = 0$ for all $\xi_0 \in A_m, \xi_0 > 0$ hence $f_1 \equiv 0$. Also, it follows that proposition 3.12 is valid for all $g_1 \in H_k^m(\Lambda, N, \overline{\psi})$. For an arbitrary Siegel modular form $g \in \mathcal{M}_k^m(N, \overline{\psi})$ let us write $g =$

g_1+h with $g_1 \in H_k^m(\Lambda, N, \overline{\psi})$ and $\langle g_1, h_1\rangle_N = 0$. Now if we combine (3.34) and (3,34a) and get the following:

$$g^\sigma = g_1^\sigma + h_1^\sigma, \quad g_1^\sigma \in H_k^m(\Lambda^\sigma, N, \overline{\psi}^\sigma), \quad \langle g_1^\sigma, h_1^\sigma\rangle_N = 0. \tag{3.41}$$

In order to obtain the above equality we used the important fact about the invariance of the subspace of all Eisenstein series in $\mathcal{M}_k^m(N,\psi)$ under the action of $\sigma \in \mathrm{Aut}(\mathbf{C})$. In turn, this property follows from the general decomposition theorem [Kl3], describing the subspace of all Eisenstein series (the orthogonal complement to the subspace of cusp forms) in terms of the Klingen – Eisenstein series, and from the invariance properties of these series under the action of complex automorphisms $\sigma \in \mathrm{Aut}(\mathbf{C})$, established by M.Harris [Har2], [Har4] (see also [Bö2], [Kur-Miz], [Miz1], [Miz2]). This last fact (stated in a more precise form) comprise the content of the Garrett's conjecture, which was proved in [Har4]. However, we do not use this fact any more and therefore will not go into detail of this interesting research. Returning to the equality (3.41) we get

$$\begin{aligned}&[\langle f^\rho, g\rangle_N \mu(\Lambda, k, \psi)^{-1}]^\sigma = [\langle f^\rho, g_1\rangle_N \mu(\Lambda, k, \psi)^{-1}]^\sigma = \\ &\langle f^{\sigma\rho}, g_1^\sigma\rangle_N \mu(\Lambda^\sigma, k, \psi^\sigma)^{-1} = \langle f^{\sigma\rho}, g^\sigma\rangle_N \mu(\Lambda^\sigma, k, \psi^\sigma)^{-1},\end{aligned}$$

To accomplish the proof of proposition 3.12 we need only to check the property (3.38) which will be now stated in a more general form.

3.14. Proposition. *Let* $\chi \bmod N$ *be an arbitrary Dirichlet character, and*

$$K^+(\xi_0, \psi, \chi) = K^+(z, k-\nu-m; \xi_0, \psi, \chi) \in \mathcal{M}_k^m(N, \overline{\psi})$$

denote a modular form in the integral representation (3.30). Then for all $\sigma \in \mathrm{Aut}(\mathbf{C})$ *there is the following relation*

$$[G(\psi\chi)^{m-1} K^+(\xi_0, \psi, \chi)]^\sigma = G(\psi^\sigma\chi^\sigma)^{m-1} K^+(\xi_0, \psi^\sigma, \chi^\sigma) \tag{3.42}$$

Proof. From the definition of the theta series we easily see that

$$\theta_{2\xi_0}^{(\nu)}(z,\chi)^\sigma = \theta_{2\xi_0}^{(\nu)}(z,\chi^\sigma).$$

Therefore it suffices to check the following property:

$$\begin{aligned}&[G(\psi\chi)^{m-1}\det\xi_0^{1/2} G^+(z,0;\chi\chi_{\xi_0}\psi, k-(m/2)-\nu, N)|W(N)]^\sigma = \\ &G(\psi^\sigma\chi^\sigma)^{m-1}((\det\xi_0^{)}1/2)^\sigma G^+(z,0;(\chi\chi_{\xi_0}\psi)^\sigma, k-(m/2)-\nu, N)|W(N),\end{aligned} \tag{3.43}$$

(in view of the equality

$$\det\xi_0^{k-(m+1)/2} = \det\xi_0^{k-(m/2)-1}, \quad m/2 \in \mathbf{Z}.)$$

According to formulas in §3 (3.53) of chapter 2, the Fourier coefficient of the series $G^+(z,0;\chi\chi_{\xi_0}\psi, k-(m/2)-\nu, N)$ by $e_m(hz)$ with $h \in A_m, h > 0$ has the form

$$(\det h)^{k-\nu-m-(1/2)} L^+(k-\nu-m, \chi\chi_h\chi_{\xi_0}\psi) M(h, \chi\chi_{\xi_0}\psi, s), \tag{3.44}$$

where

$$M(h,\chi\chi_{\xi_0}\psi,s)=\prod_{q\in P(h)} M_q(h,\chi\chi_{\xi_0}\psi(q),q^{-s})$$

is the finite Euler product (3.44) of the previous chapter, where the product is extended to all primes q in the set $P(h)$ of prime divisors of N and the elementatry divisors of the matrix h, with the property $M_q(h,t)\in\mathbf{Z}[t]$. Therefore

$$M(h,\chi\chi_{\xi_0}\psi,k-\nu-(m/2))^{\sigma}=M(h,(\chi\psi)^{\sigma}\chi_{\xi_0},k-\nu-(m/2)).$$

Now let us consider the factor $L^{+}(k-\nu-m,\chi\chi_h\chi_{\xi_0}\psi)$ and recall an elementary result on the special values of the Dirichlet L-functions [Le1], [Shi3], [Wa], [La3]. Let χ mod N be a Dirichlet character of conductor N_0, and χ_0 mod N_0 be the corresponding primitive character,

$$G(\chi)=G(\chi_0)=\sum_{x=1}^{N_0}\chi_0(x)e(x/N_0)$$

be its Gauss sum. We set, for a positive integer r

$$P(r,\chi)=G(\chi)^{-1}(2\pi i)^{-r}L(r,\chi).$$

Then for all $\sigma\in\mathrm{Aut}(\mathbf{C})$ and $\chi(-1)=(-1)^r$ we have that

$$P(r,\chi)^{\sigma}=P(r,\chi^{\sigma}) \tag{3.45}$$

If we apply this property to normalized Dirichlet L-series we see that

$$[G(\chi)^{-1}L^{+}(r,\chi)]^{\sigma}=G(\chi^{\sigma})^{-1}L^{+}(r,\chi^{\sigma})\quad (r\in\mathbf{Z},r>0) \tag{3.46}$$

$$L^{-}(r,\chi)^{\sigma}=L^{-}(r,\chi^{\sigma})\quad (r\in\mathbf{Z},r\leq 0) \tag{3.47}$$

so that for the values of the "wrong parity" the corresponding values vanish. From the basic property of Gauss sums it follows that

$$G(\chi)^{\sigma}=\chi^{\sigma}(v)^{-1}G(\chi^{\sigma})\ for\ v\in(\mathbf{Z}/N\mathbf{Z})^{\times},\ \text{such that } e(1/N)^{\sigma}=e(v/N).$$

The last property implies the useful relation:

$$\frac{G(\psi\chi)^{\sigma}}{G(\psi^{\sigma}\chi^{\sigma})}=\frac{G(\psi)^{\sigma}G(\chi)^{\sigma}}{G(\psi^{\sigma})G(\chi^{\sigma})}. \tag{3.48}$$

Let us now apply (3.46) and (3.48) to the coefficients (3.44), then we get

$$\begin{aligned}&[G(\psi\chi)^{-1}(\det\xi_0 h)^{1/2}L^{+}(k-\nu-m,\chi\chi_h\chi_{\xi_0}\psi]^{\sigma}=\\&G(\psi^{\sigma}\chi^{\sigma})^{-1}((\det\xi_0 h)^{1/2})^{\sigma}L^{+}(k-\nu-m,(\chi\psi)^{\sigma}\chi_{\xi_0}\chi_h).\end{aligned} \tag{3.49}$$

In the equality (3.49) we used the following elementary property of Gauss sums:

$$G(\chi_{\xi_0}\chi_h)\,(\det(\xi_0 h))^{1/2}\in\mathbf{Q}$$

which is due to the fact that $\chi_{\xi_0}\chi_h$ is an even quadratic character (see [B-Šaf]). To complete the proof of (3.43) and (3.44) we need the following general compatibility property of the action of Aut(**C**) and of the involution $W(N)$ (see [St2], lemma 5, and [Shi8]): for a modular form $f \in \mathcal{M}_k^m(N, \psi)$ with cyclotomic Fourier coefficients and for all $\sigma \in \mathrm{Aut}(\mathbf{C})$ we have that

$$f^\sigma|W(N) = \psi(v^m)^\sigma(f|W(N))^\sigma, \tag{3.50}$$

where $v \in (\mathbf{Z}/N\mathbf{Z})^\times$ is chosen so that $e(1/N)^\sigma = e(v/N)$. Now the proof of proposition 3.12 is finished.

3.15. Now in order to deduce theorem 3.2 on algebraicity of the normalized standard zeta function and the algebraicity property of the normalized distributions we use the already proved algebraicity properties of Fourier coefficients of the functions $K^\pm(z, s; \xi_0, \chi)|W(N)$ in the corresponding integral representation (3.28) and of the functions $F^\pm(z, \chi)$ in (3.13) and (3.14). We apply these properties in the form given below: let $k > 2m+2$, $f \in H_k^m(\Lambda, N, \psi) \subset \mathcal{S}_k^m(N, \psi)$ be a cusp form, an eigenfunction of the Hecke algebra $\mathcal{L}^m(N)$ with an eigenvalue given as the homomorphism $\Lambda : \mathcal{L}^m(N) \to \mathbf{C}$. Let us consider a linear functional

$$\mathcal{L}_f : g \mapsto \frac{\langle f^\rho, g|W(N)\rangle_N}{\langle f, f\rangle_N} \tag{3.51}$$

on the vector space $\mathcal{M}_k^m(N, \psi)$ with

$$g = \sum_{h \geq 0} b(h)e_m(hz) \in \mathcal{M}_k^m(N, \psi)$$

being its arbitrary element. Then there exist positive matrices $h_1, h_2, \cdots, h_t \in A_m$ and algebraic numbers $\alpha_1, \alpha_2, \cdots, \alpha_t \in \mathbf{Q}(f, \Lambda, \psi)$ from the field $\mathbf{Q}(f, \Lambda, \psi)$ generated by the Fourier coefficients of f and values of the homomorphism Λ and the character ψ such that for all $g \in \mathcal{M}_k^m(N, \psi)$ the linear functional is explicitly given by

$$\mathcal{L}_f(g) = \sum_i \alpha_i b(h_i). \tag{3.52}$$

Indeed, we notice that every Siegel modular form of weight $k > 2m$ is uniquely determined by its Fourier coefficients with positive matrix indices $h \in A_m$. This fact is equivalent to saying that for such a weight k there are no singular modular forms (i.e. having only Fourier coefficients with degenerate $h \in A_m, \det h = 0$), which was established by G.L.Resnikov ([Res], [Rag3]). Then proposition 3.12 implies that the number

$$\frac{\langle f^\rho, g|W(N)\rangle_N}{\mu(\Lambda, k, \psi)^{-1}} \in \mathbf{Q}(f, g, \Lambda, \psi)$$

belongs to the field $\mathbf{Q}(f, g, \Lambda, \psi)$ generated by the Fourier coefficients of the forms f and g, by the values of Λ and ψ. Moreover we have that

$$\langle f, f\rangle_N \, \mu(\Lambda, k, \psi)^{-1} \in \mathbf{Q}(f, \Lambda, \psi)^\times$$

(see the remark after proposition 3.12), and (3.52) follows. In order to prove theorem 3.2 we use proposition 3.12 and substitute in it the modular forms

$$K^{\pm}(z,s;\xi_0,\chi)|W(N) \in \mathcal{M}_k^m(N;\psi),$$

for g. This modular for has cyclotomic Fourier coefficients vanishing for degenerate matrix indices $h \in A_m$ such that the action of $\sigma \in \mathrm{Aut}(\mathbf{C})$ on them is described as in 3.14.As a consequence, we obtain also the following more explicit description of the action of $\sigma \in \mathrm{Aut}(\mathbf{C})$ on the special values in question. Put

$$\tilde{\mathcal{D}}^+(s,f,\chi) = G(\psi\chi)^{m-1}\mathcal{D}^+(s,f,\chi)\mu(\Lambda,k,\psi)^{-1},$$

$$\tilde{\mathcal{D}}^-(s,f,\chi) = G(\psi\chi)^{m}\mathcal{D}^-(s,f,\chi)\mu(\Lambda,k,\psi)^{-1}.$$

Then under the assumption of theorem 3.5 for every $\sigma \in \mathrm{Aut}(\mathbf{C})$ we have that

$$\tilde{\mathcal{D}}^{\pm}(s,f,\chi)^{\sigma} = \tilde{\mathcal{D}}^{\pm}(s,f^{\sigma},\chi^{\sigma}). \tag{3.53}$$

In particular, this relation proves the theorem 3.2. In order to deduce the algebraicity properties of proposition 3.4 for the normalized distributions we substitute in 3.12 $f_0|V(C) \in \mathcal{S}_k^m(N_0C,\psi)$ and $F^{\pm}(z,\chi)$ for f and g (see (3.13), (3.14)). In this situation the constant $\mu(\Lambda,k,\psi)$ depends only on the Λ-packet of f and it coincides with that for the original cusp form f as above.

§4. Integrality properties and congruences for the distributions

4.1. The proof of theorem 1.6 is based on a regularization of the $\overline{\mathbf{Q}}$-valued distributions

$$\mathcal{D}^+_{s,M}(s) \ \ (1 \le s \le k-\nu-m) \text{ and } \mathcal{D}^-_{s,M}(s) \ \ (1-k+\nu+m \le s \le 0)$$

from §3.

Let c be a positive integer with $(c,N_0)=1$, $c>1$ and $N_0 = 4q_0M_0^{\tilde{m}-1}C$. Then we see as in §2 that there exist distributions $\mathcal{D}_s^{c-}$, $\mathcal{D}_s^{c+}$ on $\mathbf{Z}_S^{\times}$ (*regularized distributions*), which are uniquely determined by

$$\mathcal{D}^{c-}_{s,M}(\chi) = (1-(\chi\overline{\psi})^2(c)c^{2(s-1)})\mathcal{D}^-_{s,M}(\chi), \tag{4.1}$$

$$\begin{aligned}\mathcal{D}^{c+}_{s,M}(\chi) =& C^s_{\overline{\psi}\chi}G(\overline{\psi}\chi)^{-1}(1-(\overline{\chi}\psi)^2c^{-2s})\mathcal{D}^+_{s,M}(\chi)\times \\ &\times \prod_{q|N_0}\{(1-(\overline{\psi}\chi)(q)q^{s-1})(1-(\overline{\chi}\psi)(q)q^{-s})^{-1}\},\end{aligned} \tag{4.2}$$

where $\mathcal{D}^+_{s,M}(\chi)$ and $\mathcal{D}^-_{s,M}(\chi)$ are defined by (3.8) and (3.9). Under the assumption of §1 we have that $(C,C_\chi)=1$ where C_χ is the conductor of χ. From basic elementary relations for Gauss sums it follows then that

$$\begin{aligned}\frac{C^s_{\overline{\psi}\chi}}{G(\overline{\psi}\chi)} &= \frac{C^s_{\psi}C^s_{\chi}}{G(\overline{\chi})\chi^{-1}(C_\chi)G(\psi)\psi(C_\chi)} = \\ &C(\psi)^s\chi(C_\psi)\frac{C^s_\chi}{G(\overline{\chi})}[G(\psi)\psi(C_\chi)]^{-1}.\end{aligned} \tag{4.3}$$

In order to define S-adic measures in the main theorem we put $s = 0$ and apply the embedding $i_p : \overline{\mathbf{Q}} \hookrightarrow \mathbf{C}_p$, then we get

$$\mathcal{D}^{c+} = i_p(\mathcal{D}_0^{c+}), \quad \mathcal{D}^{c-} = i_p(\mathcal{D}_0^{c-}). \tag{4.4}$$

4.2. Proposition. *(a) For all integers s with the condition $1 \le s \le k - \nu - m$ we have that the distributions $i_p(\mathcal{D}_0^{c+})$ are bounded and the following equality holds*

$$\int_{\mathbf{Z}_S^\times} \chi\, x_p^s\, d\mathcal{D}^{c+} = i_p(\mathcal{D}_{s,M}^{c+}(\chi)), \tag{4.5}$$

in which both sides vanish for $s \not\equiv \delta(\mathrm{mod}2)$.

(b) For all integers s with the condition $1 - k + \nu + m \le s \le 0$ we have that the distributions $i_p(\mathcal{D}_0^{c-})$ are bounded and the following equality holds

$$\int_{\mathbf{Z}_S^\times} \chi\, x_p^s\, d\mathcal{D}^{c-} = i_p(\mathcal{D}_{s,M}^{c-}(\chi)), \tag{4.6}$$

in which both sides vanish for $s \equiv \delta(\mathrm{mod}2)$ (Recall that

$$\nu, \delta = 0, 1 \ \ , (-1)^\nu = \chi(-1), \ \ (-1)^\delta = \psi\chi(-1)\ ).$$

4.3. *The proof* of the proposition 4.2 is based on integral representations (3.13) and (3.14). Taking into account the regularizing factors in (4.1) and (4.2) we deduce that for the corresponding values of $s \in \mathbf{Z}$ given in the proposition 4.2 the following hold

$$\langle f, f\rangle_C \mathcal{D}_{s,M}^{c\pm}(\chi) = \gamma(M')\langle f_0^\rho | V(C), F_{M'}^{c\pm}(s, \chi_M) | W(CN_0)\rangle_{CN_0}, \qquad (4.7);$$

here

$$\begin{aligned} F_{M'}^{c+}(s, \chi_M) &= \sum_{A_m \ni > 0} \ \sum_{C\hat{\xi}_0[h_1] + h_2 = M'h} d^{c+}(s, h_1, h_2) e_m(hz), \\ F_{M'}^{c-}(s, \chi_M) &= \sum_{A_m \ni > 0} \ \sum_{C\hat{\xi}_0[h_1] + h_2 = M'h} d^{c-}(s, h_1, h_2) e_m(hz) \end{aligned} \tag{4.8}$$

are modular forms from $\mathcal{M}_k^m(CN_0, \psi)$ with cyclotomic Fourier coefficients given by

$$\begin{aligned} d^{c+}(s, h_1, h_2) = {} & \chi_M(\det h_1) \det h_1^\nu \det h_2^{(2s-1)/2} P(h_2, h, s) M(h, \overline{\chi}\chi_{\xi_0}\psi, s + (m/2)) \times \\ & \times C_{\overline{\psi}\chi}^{s} G(\overline{\psi}\chi)^{-1} (1 - (\overline{\chi}\psi)^2 c^{-2s}) L_N^+(s, \omega_{\xi_0, h_2}) \times \\ & \times \prod_{q | N_0} \{(1 - (\overline{\psi}\chi)(q) q^{s-1})(1 - (\overline{\chi}\psi)(q) q^{-s})^{-1}\} \times \\ & \times (q_0 C)^{-m(2s+m)/4} \det\xi_0^{(s+k-1+\nu)/2} \det(q_0^{-1/2}\xi_0)^{(2\nu+m)/2}; \end{aligned} \tag{4.9}$$

with $\omega_{\xi_0,h_2} = \overline{\chi}\chi_{\xi_0}\psi\xi_{h_2}$ be a Dirichlet character modulo N_0;

$$\begin{aligned} d^-(s,h_1,h_2) = \chi_M(\det h_1)\det h_1^{\nu}P(h_2,h,1-s)\times \\ \times(1-(\chi\overline{\psi})^2(c)c^{2(s-1)})L_N^-(s,\overline{\chi}\chi_{\xi_0}\psi\xi_{h_2})M(h,\overline{\chi}\chi_{\xi_0}\psi,s+(m/2))\times \\ \times(q_0C)^{-m(2s+m)/4}\det\xi_0^{(s+k-1+\nu)/2}\det(q_0^{-1/2}\xi_0)^{(2\nu+m)/2}; \end{aligned} \tag{4.10}$$

here $P(v,u,s) \in \mathbf{Q}[v_{ij},u_{ij}; i \le j]$ denotes the polynomial explicitly given by the formula (4.32) of chapter 2 which is defined for all $s \in \mathbf{Z}, 1 \le s \le k-m-\nu, s \equiv \delta(\mathrm{mod} 2)$, does not depend on f, M, χ, M', and satisfies the property

$$P(v,u,s) = \det v^{(k-\nu-m-s)/2}(\mathrm{mod}(u_{ij})), \tag{4.11}$$

and where

$$M(h,\chi\chi_{\xi_0}\psi,s) = \prod_{q\in P(h)} M_q(h,\chi\chi_{\xi_0}\psi(q),q^{-s}) \tag{4.12}$$

is the finite Euler product (3.44) of chapter 2, with the product being extended to all prime numbers q in the set $P(h)$ of primes of N and the elementatry divisors of the matrix h, such that for all of these q we have $M_q(h,t) \in \mathbf{Z}[t]$. The summation in the inner sum of (4.8) is extended over all pairs (h_1,h_2) of integral matrices with the conditions

$$h_1 \in \mathrm{M}_m^+(\mathbf{Z}),\ \ h_2 > 0,\ \ h_2 \in A_m,\ \ C\hat{\xi}_0[h_1] + h_2 = M'h \tag{4.13}$$

(i.e. h_1 is a integral matrix with positive determinant, not necessarily symmetric, $h_2 \in A_m$ is a positive definite half integral matrix, and $C\hat{\xi}_0[h_1]$ denotes the matrix given by $C^t h_1 q_0 \xi^{-1} h_1 = C q_0 \xi_0^{-1}[h_1]$.

Now we notice that according to (4.9) and (4.10) the coefficients

$$d^-(s,h_1,h_2) = d^-(\chi_M; s,h_1,h_2)$$

does not depend on modulus M of the character χ_M (that is, they satisfy the compatibility criterion for distributions, see (3.4) of chapter 1) and they determine for fixed h_1,h_2 a distribution on $G_S = \mathbf{Z}_S^\times$ with values in $\mathbf{Q}^{\mathrm{ab}}$; these distributions will also be denoted by $d^-(s,h_1,h_2)$. As we soon will see these distributions turn out to be bounded measures, and the measures of proposition 4.2 will be expressed in terms of them.

4.4. Let us consider the **C**-linear functional

$$\mathcal{L}_f : g \mapsto \frac{\langle f^\rho, g|W(N)\rangle_N}{\langle f,f\rangle_N} \tag{4.14}$$

on the vector space $\mathcal{M}_k^m(N,\psi)$ defined by (3.51). From the explicit description of $\mathcal{L}_f$ given in 3.15 it follows that for an arbitrary

$$g = \sum_{h\ge 0} b(h)e_m(hz) \in \mathcal{M}_k^m(N,\psi)$$

there exist positive matrices $h_1,h_2,\cdots,h_t \in A_m$ and algebraic numbers $\alpha_1,\alpha_2,\cdots,\alpha_t \in \mathbf{Q}(f,\Lambda,\psi)$ from the field $\mathbf{Q}(f,\Lambda,\psi)$ generated by the Fourier coefficients of f and values of the homomorphism Λ and the character ψ such that for all $g \in \mathcal{M}_k^m(N,\psi)$ we have that

$$\mathcal{L}_f(g) = \sum_i \alpha_i b(h_i). \tag{4.15}$$

According to proposition 4.5 of chapter 2 the values of the distributions $\mathcal{D}_s^{c\pm}$ can be represented in terms of the functional $\mathcal{L}$ as follows

$$\mathcal{D}_s^{c\pm}(\chi) = \gamma(M')\mathcal{L}(F_{M'}^{c\pm}(s,\chi_M)). \tag{4.16}$$

Combining (4.15) and (4.16) we see that

$$\mathcal{D}_s^{c\pm}(\chi) = \sum_i \alpha_i\, v^{c\pm}(M'h^{(i)}, s, \chi_M), \tag{4.17}$$

where

$$v^{c\pm}(M'h^{(i)}, s, \chi_M) = \sum_{C\hat{\xi}_0[h_1]+h_2=M'h} d^{c\pm}(s,h_1,h_2) \tag{4.17a}$$

are the Fourier coefficients of the functions $F_{M'}^{c\pm}(s,\chi_M)$. Therefore the statements (a) and (b) of proposition 4.2 under the assumptions of the main theorem are equivalent to the corresponding statements about the distributions $d^{c\pm}(s,h_1,h_2)$. Indeed, the value of $|i_p(\gamma(M')|_p$ does not change if we shall vary M': in view of the definition (3.10) we have that

$$\gamma(M') = 2^{m(2k-2-m-\kappa)} i^{-m(k-(m/2)-\nu)} a_0(\xi_0)^{-1} \alpha_0(M_0^{\tilde{m}-1}M'^{-1})(CM_0^{\tilde{m}-1})^{(k-1-m)/2},$$

and if we take into account the condition $|i_p(\alpha_0(M_0))|_p = 1$ satisfied by the p-ordinary form f, we see that the denominators in the linear combination (4.17) are uniformly bounded for varying χ_M and M'. Below in 4.6 is given a more precise argument based on the generalized Kummer congruences. However we show first that the distributions $d^{c\pm}(s,h_1,h_2)$ essentially reduce to the S-adic Mazur measure for the Kubota-Leopoldt zeta function (see 4.7 of chapter 1 and [Ku-Le], [Le2], [Maz-SD], [Man4], [Man6], [Wa]) whose properties are recalled in the following 4.5.

4.5. Let ω mod A be a fixed primitive Dirichlet character such that $(A, M_0) = 1$ with $M_0 = \prod_{q\in S} q$. Put $\overline{S} = S \cup S(A)$, $\overline{M} = \prod_{q\in\overline{S}} q$. Then for any positive integer c with $(c,\overline{M}) = 1$, $c > 1$ there exist $\mathbf{C}_p$-measures $\mu^+(c,\omega), \mu^-(c,\omega)$ on $\mathbf{Z}_S^\times$ which are uniquely determined by the following conditions: for $s \in \mathbf{Z}, s > 0$

$$\begin{aligned} i_p^{-1}\left(\int_{\mathbf{Z}_S^\times} \chi\, x_p^s\, d\mu^+(c,\omega)\right) &= (1-\overline{\chi}\omega(c)c^{-s})\frac{C_{\omega\overline{\chi}}}{G(\omega\overline{\chi})}\times \\ \times \prod_{q\in S\setminus S(\chi)} &\{(1-\chi\overline{\omega}(q)q^{s-1})/(1-\overline{\chi}\omega(q)q^{-s})\} L_{M_0}^+(s,\overline{\chi}\omega), \end{aligned} \tag{4.18}$$

and for $s \in \mathbf{Z}, s \le 0$

$$i_p^{-1}\left(\int_{\mathbf{Z}_S^\times} \chi\, x_p^s\, d\mu^-(c,\omega)\right) = (1-\chi\overline{\omega}(c)c^{s-1})L_{M_0}^-(s,\overline{\chi}\omega), \tag{4.19}$$

where

$$L^{+}_{M_0}(s,\overline{\chi}\omega) = L_{\overline{M}}(s,\overline{\chi}\omega)\frac{2i^{\delta}\Gamma(s)\cos(\pi(s-\delta)/2)}{(2\pi)^s}, \tag{4.20}$$

$$L^{+}_{M_0}(s,\overline{\chi}\omega) = L_{\overline{M}}(s,\overline{\chi}\omega) \tag{4.21}$$

are the normalized Dirichlet L-functions with $\delta = 0, 1$, $(-1)^{\delta} = \overline{\chi}\omega(-1)$. The functions (4.18) and (4.19) satisfy the following functional equation

$$L_{M_0}(1-s,\chi\overline{\omega}) = \prod_{q\in S\setminus S(\chi)} \{(1-\chi\overline{\omega}(q)q^{s-1})/(1-\overline{\chi}\omega(q)q^{-s})\}L^{+}_{M_0}(s,\overline{\chi}\omega). \tag{4.22}$$

The properties (4.18) - (4.22) follow easily if we remember the definition of the $\overline{S}$-adic Mazur measure μ^c on $\mathbf{Z}^{\times}_{\overline{S}}$ for which (4.18) and (4.19) are given by the equalities

$$\int_{\mathbf{Z}^{\times}_{S}} x\, d\mu^{-}(c,\omega) = \int_{\mathbf{Z}^{\times}_{\overline{S}}} x^{-1}\omega\, d\mu^c,$$

$$\int_{\mathbf{Z}^{\times}_{S}} d\mu^{+}(c,\omega) = \int_{\mathbf{Z}^{\times}_{\overline{S}}} x\, x_p^{-1}\omega^{-1}\, d\mu^c,$$

in which $x \in X_S$, and $X_{\overline{S}}$ is regarded as a subgroup of $X_{\overline{S}}$.

4.6. To accomplish the proof of proposition 4.2 we recall *the abstract Kummer congruences* (see proposition 3.3 of chapter 1, and [Ka3], p.258) which gives a criterion of boundness for p-adic valued distributions as follows. Let $\{f_i\}$ be a system of continuous functions $f_i \in \mathcal{C}(Y,\mathcal{O}_p)$ in the ring $\mathcal{C}(Y,\mathcal{O}_p)$ of all continuous functions over the compact totally disconnected group $Y = \mathbf{Z}^{\times}_{S}$ with values in the ring of integers $\mathcal{O}_p$ of $\mathbf{C}_p$ such that the $\mathbf{C}_p$-linear span of $\{f_i\}$ is dense in $\mathcal{C}(Y,\mathbf{C}_p)$. Let also $\{a_i\}$ be a system of elements $a_i \in \mathcal{O}_p$. Then existence of a $\mathcal{O}_p$-valued measure μ on Y with the property

$$\int_Y f_i\, d\mu = a_i$$

is equivalent to the following congruences: for an arbitrary choice of elements $b_i \in \mathbf{C}_p$ almost all of which vanish

$$\sum_i b_i f_i(y) \in p^n\mathcal{O}_p \text{ for all } y \in Y \text{ implies } \sum_i b_i a_i \in p^n\mathcal{O}_p.$$

In our situation we take the functions f_i to be all of the functions of the form χx_p^s with s as in proposition 4.2, and $\chi \in X_S^{\mathrm{tors}}$. This system of functions obviously has the dense $\mathbf{C}_p$-linear span. For any finite number of Dirichlet characters $\chi \in X_S^{\mathrm{tors}}$ we choose an integer M and a sufficiently large integer M' such that each of these characters is defined modulo M and the formula of the proposition 3.6 holds for the values of distributions $\mathcal{D}^{c\pm}_{s,M}$. Then we apply the functional $\mathcal{L}$ and use the equality (4.17). As it was mentioned above, the coefficients of the linear combination in (4.17) are p-adically bounded for varying M' and χ hence the proof of the abstract Kummer congruences for the numbers $\mathcal{D}^{c\pm}_{s,M}$ is reduced to verifying the correspondent congruences for the numbers $v^{c\pm}(M'h,s,\chi)$ (see (4.17a)). In turn, for this purpose we fix h_1 and use the formulas (4.9) and (4.10). It is seen from the relation $C\hat{\xi}_0[h_1] + h_2 = M'h$ in (4.8) that the following congruence holds:

$$\det h_2 \equiv (-C)^m \det\hat{\xi}_0 \det h_1^2 \ (\mathrm{mod} M') \tag{4.23}$$

with $\det\hat{\xi}_0 = q_0^m \det\xi_0^{-1}$. If we then use the property (4.11) of the polynomial $P(v, u, s)$ and (4.23) then we get for the factor $P(h_2, h, s)$ in (4.9) and (4.10) the following congruence:

$$P(h_2, h, s) \equiv [(-Cq_0)^m \det\xi_0{}^{-1} \det h_1^2]^{(k-\nu-m-s)/2} (\mathrm{mod} M') \tag{4.24}$$

where $1 \leq s \leq k - \nu - m,\ s \equiv \delta(\mathrm{mod} 2)$.

It can happen that the congruence (4.24) is valid only modulo a slightly smaller positive integer than M' (i.e. obtained from M' by dividing it by a divisor independent on the choice of M'). However, with the growing M' we may ignore this divisor when we multiply the numbers (4.9) and (4.10) by a suitable positive integer independent of M'.

Recall that in formulas (4.9) and (4.10) we used the notation $\omega_{\xi_0, h_0} = \omega$ for the primitive Dirichlet character associated with $\chi_{\xi_0}\chi_{h_2}\psi$, and if $\det h_2 \det(\hat{\xi}_0) = a^2 t$ with a square free integer t then we have that $\omega = \chi_t \psi$ where χ_t is the primitive Dirichlet character associated with the quadratic field $\mathbf{Q}(\sqrt{t})$. The congruence (4.23) imply in particular that $t \equiv 1(\mathrm{mod} 4)$ hence the conductor of χ_t is equal to t. Indeed if $q|M'$ (e.g. $q = 2$) then according to (4.23) we get

$$\left(\frac{\det h_2 \det\hat{\xi}_0}{q}\right) = \left(\frac{\det h_1^2 \det\hat{\xi}_0^2 (-C)^m}{q}\right),$$

in view of the parity of m we have $\omega(q) = \psi(q)$ for $q|M_0\ \ (q \in S)$ and $\omega^2 = \psi^2$, and also that $(t, M_0) = 1$.

Now we compare the formulas (4.9) and (4.10) with the corresponding formulas (4.18) and (4.19) for the Dirichlet L-series, and take into account that

$$\begin{aligned} &C_{\omega\overline{\chi}} = tC_{\psi\overline{\chi}},\ \omega = \chi_t\psi,\ \chi_t(C_{\psi\overline{\chi}}) = 1, \\ &G(\omega\overline{\chi}) = G(\chi_t\psi\overline{\chi}) = \chi_t(C_{\psi\overline{\chi}}\psi\overline{\chi}(t)G(\chi_t)G(\psi\overline{\chi}) = \psi\overline{\chi}(t)\sqrt(t)G(\psi\overline{\chi}), \end{aligned} \tag{4.25}$$

$$(1 - (\overline{\chi}\psi)(c)^{-2s}) = (1 - (\omega\overline{\chi}c^{-s})(1 + (\omega\overline{\chi}c^{-s}). \tag{4.26}$$

Next let us apply the embedding i_p to (4.9) and (4.10) keeping in mind (4.25) and (4.26), then (4.9) and (4.9) take the following form

$$\begin{aligned} &i_p\left(d^{c+}(s, h_1, h_2)\right) = \\ &\frac{\psi(t)\sqrt{t}}{\chi(t)t^s}(1 + (\omega\overline{\chi}c^{-s}) \int_{\mathbf{Z}_S^\times} \chi\, x_p^s\, d\mu^-(c, \omega) \times \\ &\times \chi_M(\det h_1) \det h_1^\nu \det h_2^{(2s-1)/2} P(h_2, h, s) M(h, \overline{\chi}\chi_{\xi_0}\psi, s + (m/2)) \times \\ &\times (q_0 C)^{-m(2s+m)/4} \det\xi_0^{(s+k-1+\nu)/2} \det(q_0^{-1/2}\xi_0)^{(2\nu+m)/2}; \end{aligned} \tag{4.27}$$

for $1 \le s \le k - \nu - m$,

$$
\begin{aligned}
&i_p\left(d^-(s,h_1,h_2)\right) = \\
&(1+(\omega\overline{\chi}c^{-s}))\int_{\mathbf{Z}_S^\times} \chi\, x_p^s\, d\mu^-(c,\omega)\times \\
&\times \chi_M(\det h_1)\det h_1^{\nu} P(h_2,h,1-s) M(h,\overline{\chi}\chi_{\xi_0}\psi, s+(m/2))\times \\
&\times (q_0C)^{-m(2s+m)/4}\det\xi_0^{(s+k-1+\nu)/2}\det(q_0^{-1/2}\xi_0)^{(2\nu+m)/2};
\end{aligned}
\tag{4.28}
$$

for $1-k+\nu+m \le s \le 0$.

Notice that the finite Euler product $M(h,\overline{\chi}\chi_{\xi_0}\psi, s+(m/2))$ is a finite linear combination of terms of the type $\overline{\chi}(b)b^{-s} = (\chi(b)b^s)^{-1}$ (with $(b,M')=1$) in which coefficients are algebraic integers independent of χ. Bearing in mind the congruences (4.23) we obtain from (4.27) and (4.28) that for the corresponding values of s the expression for $v^{c\pm}(M'h,s,\chi)$ mod M' takes the form of the following linear combination:

$$
\begin{aligned}
&\sum_i A_i\chi(y_i)y_i^s \int_{\mathbf{Z}_S^\times} \chi\, x_p^s\, d\mu^\pm(c,\omega) = \\
&\sum_i A_i \int_{\mathbf{Z}_S^\times} (\chi\, x_p^s)(y_i y)\, d\mu^\pm(c,\omega)(y) \quad (y, y_i \in \mathbf{Z}_S^\times),
\end{aligned}
\tag{4.29}
$$

with uniformly p-adically bounded algebraic coefficients $A_i \in i_p(\mathbf{Q}^{ab})$. It remains to notice that the abstract Kummer congruences are tautologically satisfied by the expressions of the type (4.29) (which obviously satisfy the identities of the form (4.5) and (4.6)). Hence, the corresponding statements also hold for the distributions $\mathcal{D}_s^{c\pm}$, and the proposition 4.2 follows.

Finally, we get the proof of the main theorem as a combination of all of the results obtained above: proposition 4.2, the relation between the values of the normalized distributions and the values of the normalized standard zeta-functions (proposition 2.3), the definitions of the normalizing factors (3.2), (3.3), (3.8), (3.9) and of the regularizing factors (4.1), (4.2).

Chapter 4. Non-Archimedean convolutions of Hilbert modular forms.

§0. Introduction

0.1. Now let p be a prime number and S a finite set of primes containing p. In this chapter we consider convolutions of Hilbert modular forms and construct their S-adic analogues; they correspond to certain automorphic forms on the group $G = \mathrm{GL}_2 \times \mathrm{GL}_2$ over a totally real field F and have the form

$$L(s, \mathbf{f}, \mathbf{g}) = \sum_{\mathfrak{n}} C(\mathfrak{n}, \mathbf{f})\, C(\mathfrak{n}, \mathbf{g}) \mathcal{N}(\mathfrak{n})^{-s}, \tag{0.1}$$

where $\mathbf{f}$, $\mathbf{g}$ are Hilbert automorphic forms of "holomorphic type" over F, and $C(\mathfrak{n}, \mathbf{f})$, $C(\mathfrak{n}, \mathbf{g})$ are their normalized Fourier coefficients (indexed by integral ideals $\mathfrak{n}$ of the maximal order $\mathcal{O}_F \subset F$). We regard $\mathbf{f}$, $\mathbf{g}$ as functions on the adelic group $G_{\mathbf{A}} = \mathrm{GL}_2(\mathbf{A}_F)$, where $\mathbf{A}_F$ is the ring of adeles of F and we suppose that $\mathbf{f}$ is a primitive cusp form of scalar weight $k \geq 2$, conductor $\mathfrak{c}(\mathbf{f}) \subset \mathcal{O}_F$, and charachter ψ and $\mathbf{g}$ a primitive cusp form of weight $l < k$, conductor $\mathfrak{c}(\mathbf{g})$, and character ω (here $\psi, \omega : \mathbf{A}_F^{\times} \to \mathbf{C}^{\times}$ are Hecke characters of finite order). The non-Archimedean construction is based on the algebraicity properties of the special values of the function $L(s, \mathbf{f}, \mathbf{g})$ at the points $s = l, \cdots, k-1$ (up to some constant which can be expressed in terms of the Petersson inner product $\langle \mathbf{f}, \mathbf{f} \rangle$ of the automorphic form $\mathbf{f}$ [Shi6]). Our theorem on non-Archimedean interpolation is equivalent to certain generalized Kummer congruences for these special values.

0.2. We need some more notations for the precise formulation of the result (in a simplified form). Let ψ^*, ω^* be the characters of the ideal group of F which are associated with ψ, ω and let

$$L_{\mathfrak{c}}(s, \psi\omega) = \sum_{\mathfrak{n} + \mathfrak{c} = \mathcal{O}_F} \psi^*(\mathfrak{n})\omega^*(\mathfrak{n})\mathcal{N}(\mathfrak{n})^{-s} = \prod_{\mathfrak{p} + \mathfrak{c} = \mathcal{O}_F} (1 - \psi^*(\mathfrak{p})\omega^*(\mathfrak{p})\mathcal{N}(\mathfrak{p})^{-s})^{-1} \tag{0.2}$$

be the correspoding Hecke L-function (here $\mathfrak{c} = \mathfrak{c}(\mathbf{f})\mathfrak{c}(\mathbf{g})$). We now define the normalized zeta function by setting

$$\Psi(s, \mathbf{f}, \mathbf{g}) = \gamma_n(s) L_{\mathfrak{c}}(2s + 2 - k - l, \psi\omega) L(s, \mathbf{f}, \mathbf{g})$$

where $n = [F : \mathbf{Q}]$ is the degree of F,

$$\gamma(s) = (2\pi)^{-2ns}\Gamma(s)^n \Gamma(s + 1 - l)^n$$

is the gamma-factor. Then the function $\Psi(s, \mathbf{f}, \mathbf{g})$ admits a holomorphic analytic continuation onto the entire comlex plane, and it satisfies a certain functional equation [Ja], [Shi6]. If we put $\Omega(\mathbf{f}) = \langle \mathbf{f}, \mathbf{f} \rangle_{\mathfrak{c}(\mathbf{f})}$, then the number

$$\frac{\Psi(l + r, \mathbf{f}, \mathbf{g})}{(2\pi i)^{n(1-l)}\Omega(\mathbf{f})} \text{ is algebraic for all integers } r \text{ with } 0 \leq r \leq k - l - 1. \tag{0.3}$$

(see [Shi6], theorem 4.2; the precise definition of $\langle \mathbf{f},\ \mathbf{f}\rangle_{c(\mathbf{f})}$ is given in §1 below). For the non-Archimedean construction we consider the S-adic completion

$$\mathcal{O}_S = \prod_{q \in S} (\mathcal{O}_F \otimes \mathbf{Z}_q) = \prod_{\mathfrak{p} | q \in S} \mathcal{O}_\mathfrak{p}$$

of the ring $\mathcal{O}_F$.

We set

$$S_F = \{\mathfrak{p} \mid \mathfrak{p} \text{ divides } q \in S\}, \quad \mathfrak{m}_0 = \prod \mathfrak{p} \text{ (over all } \mathfrak{p} \in S_F),$$

and let $\mathrm{Gal}_S = \mathrm{Gal}(F(S)/F)$ denote the Galois group of the maximal abelian extension of F unramified outside S and ∞.

The domain of definition of the non-Archimedean L-functions is the p-adic analytic Lie group

$$\mathcal{X}_S = \mathrm{Hom}_{\mathbf{contin}}(\mathrm{Gal}_S, \mathbf{C}_p^\times)$$

of all continuouos p-adic characters of the Galois group Gal_S ($\mathbf{C}_p$ is the Tate field). Elements of finite order $\chi \in \mathcal{X}_S$ can be identified with those Hecke characters of finite order whose conductors contain only prime divisors in S_F; this identification uses the map

$$\chi : \mathbf{A}_F^\times \overset{\mathrm{CFT}}{\longrightarrow} \mathrm{Gal}_S \to \overline{\mathbf{Q}}^\times \overset{i_p}{\to} \mathbf{C}_p^\times,$$

where CTF is the homomorphism of class field theory.

We shall use the same symbol χ to denote both Hecke character and the corresponding element of $\mathcal{X}_S$. Since $\mathbf{Q}(S) \subset F(S)$, the restriction of Galois automorphisms to $\mathbf{Q}(S)$ determines a natural homomorphism

$$\mathcal{N} : \mathrm{Gal}_S \to \mathrm{Gal}(\mathbf{Q}(S)/\mathbf{Q}) \cong \mathbf{Z}_S^\times = \prod_{q \in S} \mathbf{Z}_q^\times. \tag{0.4}$$

We shall let $\mathcal{N}x_p$ denote the composition of this homomorphism with the natural projection $\prod_q \mathbf{Z}_q^\times \to \mathbf{Z}_p^\times$ and the inclusion $\mathbf{Z}_p^\times \subset \mathbf{C}_p^\times$.

Starting from the property (0.3), we construct a bounded p-adic analytic function $\Psi_S : \mathcal{X}_S \to \mathbf{C}_p$, $\Psi_S = \Psi_S(\mathbf{f}, \mathbf{g})$ whose values at the characters $x = \chi \mathcal{N} x_p^r$ coincide (up to an elementary factor) with the number of the form

$$i_p\left[\frac{\Psi(l + r, \mathbf{f}, \mathbf{g}^\rho(\bar{\chi}))}{(-2\pi i)^{n(1-l)}\langle \mathbf{f},\ \mathbf{f}\rangle}\right].$$

As in chapter 3, we make an important assumption that that the cusp form $\mathbf{f}$ is *p-ordinary*, i.e. for the fixed embedding $\overline{\mathbf{Q}} \hookrightarrow \mathbf{C}_p$ and for all $\mathfrak{p} \mid p$ there exists a root $\alpha(\mathfrak{p})$ of Hecke $\mathfrak{p}$-polynomial of $\mathbf{f}$ such that $| i_p(\alpha(\mathfrak{p})) |_p = 1$. We then fix the roots $\alpha(\mathfrak{p})$ and extend the definition of $\alpha(\mathfrak{m})$ to all integral ideals $\mathfrak{m} \subset \mathcal{O}_F$ by multiplicativity.

0.3. Theorem *(On non-Archimedean convolutions of Hilbert modular forms) Under the above notations and assumptions there exists a bounded $\mathbf{C}_p$-analytic function*

$\Psi_S : \mathcal{X}_S \to \mathbf{C}_p$, which is uniquely determined by the condition: for all Hecke characters of finite order $\chi \in \mathcal{X}_S^{\text{tors}}$ the following equality holds

$$\Psi_S(\chi) = i_p\Big[D_F^{2l}\omega^*(\mathfrak{m})\frac{\tau(\chi)^2\mathcal{N}\mathfrak{m}^{l-1}}{\alpha(\mathfrak{m})^2}\frac{\Psi(l,\mathbf{f},\mathbf{g}^\rho(\bar{\chi}))}{(-2\pi i)^{n(1-l)}\langle\mathbf{f},\ \mathbf{f}\rangle}\Big],$$

where D_F is the discriminant of F, $\tau(\chi)$is the Gauss sum of χ, and $\mathbf{g}^\rho(\chi)$ the cusp form obtained from $\mathbf{g}$ by complex conjugation of its Fourier coefficients and by twisting it then with the character χ.

0.4. We remark that a similar result is valid for the special values $\Psi(l+r,\mathbf{f},\mathbf{g})$ with $r = 1,\cdots,k-l$, if we replace $\chi \in \mathcal{X}_S$ by $\chi\mathcal{N}x_p^r \in \mathcal{X}_S$ (see the Main theorem in §2 for the precise statement).

Also, this construction was recently extended by My Vinh Quang (Moscow University) to the *non-p-ordinary*, i.e. *supersingular* case, when $|\ i_p(\alpha(\mathfrak{p})\ |_p < 1$ for all $\mathfrak{p} \mid p$. In this situation the functions Ψ_S are also uniquely determined by the above condition provided that they have only a prescribed logarithmic growth on $\mathcal{X}_S$ (see [V1]).

0.5. The essential property of the convolution

$$L(s,\mathbf{f},\mathbf{g}(\chi)) = \sum_{\mathfrak{n}}\chi^*(\mathfrak{n})C(\mathfrak{n},\mathbf{f})\,C(\mathfrak{n},\mathbf{g})\mathcal{N}(\mathfrak{n})^{-s}$$

is the following Euler product decomposition

$$\begin{aligned} &L_{\mathfrak{c}}(2s+2-k-l,\psi\omega\chi^2)L(s,\mathbf{f},\mathbf{g}(\chi)) = \\ &\prod_{\mathfrak{q}}(1-\chi^*(\mathfrak{q})\alpha(\mathfrak{q})\beta(\mathfrak{q})\mathcal{N}(\mathfrak{q})^{-s})(1-\chi^*(\mathfrak{q})\alpha(\mathfrak{q})\beta'(\mathfrak{q})\mathcal{N}(\mathfrak{q})^{-s})\times \\ &\times(1-\chi^*(\mathfrak{q})\alpha'(\mathfrak{q})\beta(\mathfrak{q})\mathcal{N}(\mathfrak{q})^{-s})(1-\chi^*(\mathfrak{q})\alpha'(\mathfrak{q})\beta'(\mathfrak{q})\mathcal{N}(\mathfrak{q})^{-s}), \end{aligned} \tag{0.5}$$

where the numbers $\alpha(\mathfrak{q})$, $\alpha'(\mathfrak{q})$, $\beta(\mathfrak{q})$, and $\beta'(\mathfrak{q})$ are roots of the Hecke polynomials

$$X^2 - C(\mathfrak{q},\mathbf{f})X + \psi^*(\mathfrak{q})\mathcal{N}(\mathfrak{q})^{k-1} = (X-\alpha(\mathfrak{q}))(X-\alpha'(\mathfrak{q})),$$

and

$$X^2 - C(\mathfrak{q},\mathbf{g})X + \omega^*(\mathfrak{q})\mathcal{N}(\mathfrak{q})^{l-1} = (X-\beta(\mathfrak{q}))(X-\beta'(\mathfrak{q})).$$

The decomposition (0.5) is not difficult to deduce from the following elementary lemma on rational functions, applied to each of the Euler $\mathfrak{q}$-factors: if

$$\sum_{i=0}^{\infty}A_iX^i = \frac{1}{(1-\alpha X)(1-\alpha' X)},\qquad \sum_{i=0}^{\infty}B_iX^i = \frac{1}{(1-\beta X)(1-\beta' X)},$$

then

$$\sum_{i=0}^{\infty}A_iB_iX^i = \frac{1-\alpha\alpha'\beta\beta'X^2}{(1-\alpha\beta X)(1-\alpha\beta' X)(1-\alpha'\beta X)(1-\alpha'\beta' X)}. \tag{0.6}$$

0.6. Content of the chapter. We recall in §1 some basic facts about Hilbert modular forms and their Fourier coefficients. The precise formulation of the main result of the chapter is given in §2. The construction of the complex valued distributions $\tilde{\mu}_s$ on the Galois group Gal_S in §3 is the most important ingredient of the proof. We establish algebraicity properties of these distributions in §4 by means of the Rankin – Shimura integral representation for $s = l, \cdots, k-1$. When we apply the embedding $i_p : \overline{\mathbf{Q}} \hookrightarrow \mathbf{C}_p$, these values become p-adic numbers and they determine certain p-adic measures on $\mathcal{X}_S$; then the function Ψ_S of the Main theorem is constructed as the non-Archimedean Mellin transform of the p-adic measure $i_p(\tilde{\mu}_l)$.

Notations

Let the symbols

$$\mathcal{O}_F,\ I,\ F_{\mathbf{A}},\ F_{\mathbf{A}}^{\times},\ \mathfrak{d} \subset \mathcal{O}_F,\ D_F = \mathcal{N}(\mathfrak{d})$$

denote, respectively, the maximal order, the group of fractional ideals, the ring of adeles, the group of ideles, the different and the discriminant of a totally real field F of degree n over $\mathbf{Q}$. Let $\Sigma = \Sigma_\infty \cup \Sigma_0$ denote the set of places (i.e. eqivalence classes of valuations) of F where $\Sigma_\infty = \{\infty_1, \cdots, \infty_n\}$ are the Archimedean places, $\Sigma_0 = \{\mathfrak{p} = \mathfrak{p}_v \subset \mathcal{O}_F\}$ finite (non-Archimedean) places, so that if $|\cdot|_v$ are the corresponding normalized valuations then the product formula holds

$$\prod_v |x|_v \quad = \quad 1 \quad (x \in F^\times, \quad v \in \Sigma).$$

The Archimedean places are induced by the real embeddings of $F : x \mapsto x^{(\nu)} \in \mathbf{R}$ $(\nu = 1, \cdots, n)$. An element $x \in F^\times$ is called totally positive ($x \gg 0$) if one has $x^{(\nu)} > 0$ for all ν and let $F_+^\times$ denote the multiplicative group of all totally positive elements of F. We put also $F_\infty = F \otimes_{\mathbf{Q}} \mathbf{R} \cong \mathbf{R}^n \subset F_{\mathbf{A}}$, and let $\hat{F} \cong \hat{\mathcal{O}}_F \otimes_{\mathbf{Z}} \mathbf{Q} \subset F_{\mathbf{A}}$ be the subring of finite adeles where $\hat{\mathcal{O}}_F$ is the profinite completion of the ring $\mathcal{O}_F$ (with respect to all its ideals). Then $F_{\mathbf{A}} = F_\infty \oplus \hat{F}$, and for an adele $x = (x_v)_{v \in \Sigma}$ we write $x = x_\infty + x_0$ where $x_\infty \in F_\infty$, $x_0 \in \hat{F}$. On the other hand there is the decomposition $F_{\mathbf{A}}^\times = F_\infty^\times \times \hat{F}^\times$ and we shall allow ourselves convenient abuse of notation by writing $y = y_\infty \cdot y_0$ with $y_\infty \in F_\infty^\times$, $y_0 \in \hat{F}^\times$. For the idele $y \in F_{\mathbf{A}}^\times$ let the symbol $\tilde{y} \in I$ denote the fractional ideal associated with y (so that $\tilde{y}\widehat{\mathcal{O}}_F \;=\; y_0\widehat{\mathcal{O}}_F$).

§1. Hilbert modular forms

1.1. The group. We shall regard the group $\mathrm{GL}_2(F)$ as the group $G_{\mathbf{Q}}$ of all $\mathbf{Q}$-rational points of a certain $\mathbf{Q}$ -subgroup $G \subset \mathrm{GL}_{2n}$. Then the adelization $G_{\mathbf{A}} = G(\mathbf{A})$ can be identified with the product

$$\mathrm{GL}_2(F_{\mathbf{A}}) \cong G_\infty \times G_{\widehat{\mathbf{Q}}},$$

where

$$G_\infty = \mathrm{GL}_2(F_\infty) \cong \mathrm{GL}_2(\mathbf{R})^n, \quad G_{\widehat{\mathbf{Q}}} = \mathrm{GL}_2(\widehat{F}).$$

The subgroup

$$G_\infty^+ \cong \mathrm{GL}_2^+(\mathbf{R})^n \subset G_\infty$$

consists of all elements

$$\alpha = (\alpha_1, \cdots, \alpha_n), \quad \alpha_\nu = \begin{pmatrix} \alpha_\nu & \beta_\nu \\ \gamma_\nu & \delta_\nu \end{pmatrix},$$

such that $\det\alpha_\nu > 0$, $\nu = 1, \cdots, n$. Every element $\alpha \in G_\infty^+$ acts on the product $\mathfrak{H}^n$ of the n copies of the upper half planes according to the formula

$$\alpha(z_1, \cdots, z_n) = (\alpha_1(z_1), \cdots, \alpha_n(z_n)),$$

where

$$\alpha_\nu(z_\nu) \quad = \quad (a_\nu z_\nu + b_\nu)/(c_\nu z_\nu + d_\nu).$$

For $z = (z_1, \cdots, z_n)$ we put $\{z\} = z_1 + \cdots + z_n$ and $e_F(z) = e(\{z\})$, with $e(x) = \exp(2\pi i x)$. Let $\mathbf{i} = (i, \cdots, i) \in \mathfrak{H}^n$, then

$$\{\alpha \in G_\infty^+ | \alpha(\mathbf{i}) = \mathbf{i}\}/\mathbf{R}_+^\times \cong \mathrm{SO}(2)^n$$

is a maximal compact subgroup in $G_\infty^+/\mathbf{R}_+^\times$. For $\alpha \in G_\infty^+$, an integer k and an arbitrary function $f : \mathfrak{H}^n \to \mathbf{C}$ we use the notation

$$(f|_k\alpha)(z) = \mathcal{N}(cz+d)^{-k} f(\alpha(z)) \mathcal{N}\det(\alpha)^{k/2},$$

with $\mathcal{N}(z)^k = z_1^k \cdots z_n^k$. Let $\mathfrak{c} \subset \mathcal{O}_F$ be an integral ideal, $\mathfrak{c}_\mathfrak{p} = \mathfrak{c}\mathcal{O}_\mathfrak{p}$ its $\mathfrak{p}$-part, $\mathfrak{d}_\mathfrak{p} = \mathfrak{d}\mathcal{O}_\mathfrak{p}$ the local different. We shall need the open subgroups $W = W_\mathfrak{c} \subset G_\mathbf{A}$ defined by

$$\begin{aligned} &W = G_\infty^+ \times \prod_\mathfrak{p} W(\mathfrak{p}), \\ &W(\mathfrak{p}) = \\ &\left\{ \begin{pmatrix} a & b \\ c & d \end{pmatrix} \in \mathrm{GL}_2(F_\mathfrak{p}) | b \in \mathfrak{d}_\mathfrak{p}^{-1}, c \in \mathfrak{d}_\mathfrak{p}\mathfrak{c}_\mathfrak{p},\ a, d \in \mathcal{O}_\mathfrak{p}, ad - bc \in \mathcal{O}_\mathfrak{p}^\times \right\}. \end{aligned} \tag{1.1}$$

Let $h = |\widetilde{Cl_F}|$ be the number of ideal classes of F (in the narrow sense),

$$\widetilde{Cl_F} = I/\{(x) | x \in F_+^\times\},$$

and let us choose the ideles $t_1, \cdots, t_h$ so that $\tilde{t}_\lambda \subset \mathcal{O}_F$ form a complete system of representatives for $\widetilde{Cl_F}$, $(t_\lambda)_\infty = 1$ and $\tilde{t}_\lambda + \mathfrak{m}_0 = \mathcal{O}_F$ ($\lambda = 1, \cdots, h$, $\mathfrak{m}_0 = \prod_{\mathfrak{q} \in S_F} \mathfrak{q}$). If we put $x_\lambda = \begin{pmatrix} 1 & 0 \\ 0 & t_\lambda \end{pmatrix}$ then there is the following decomposition into a disjoint union ("the approximation theorem"):

$$G_\mathbf{A} = \cup_\lambda G_\mathbf{Q} x_\lambda W = \cup_\lambda G_\mathbf{Q} x_\lambda^{-\iota} W, \tag{1.2}$$

where $x_\lambda^{-\iota} = \begin{pmatrix} t_\lambda^{-1} & 0 \\ 0 & 1 \end{pmatrix}$, ι denotes the involution given by

$$\begin{pmatrix} a & b \\ c & d \end{pmatrix}^\iota \quad = \quad \begin{pmatrix} d & -b \\ -c & a \end{pmatrix}$$

(see [Shi6], p.647).

1.2. Definition of Hilbert automorphic forms of weight k and level $\mathfrak{c} \subset \mathcal{O}_F$ with a Hecke character ψ of finite order. By a *Hilbert automorphic form* of weight k, level $\mathfrak{c} \subset \mathcal{O}_F$, and Hecke character ψ we mean a function $\mathbf{f} : G_{\mathbf{A}} \to \mathbf{C}$ satisfying the following conditions (1.3) – (1.5):

$$\begin{aligned} &\mathbf{f}(s\alpha x) = \psi(s)\mathbf{f}(x) \text{ for all } x \in G_{\mathbf{A}} \\ &\text{for } s \in F_{\mathbf{A}}^{\times} \text{ (the center of } G_{\mathbf{A}}\text{), and } \alpha \in G_{\mathbf{Q}}. \end{aligned} \tag{1.3}$$

If we let $\psi_0 : (\mathcal{O}_F/\mathfrak{c})^{\times} \to \mathbf{C}^{\times}$ denote the $\mathfrak{c}$-part of the character ψ, and then extend the definition of ψ over W by the formula

$$\psi\left(\begin{pmatrix} a & b \\ c & d \end{pmatrix}\right) = \psi_0(a_{\mathfrak{c}} \bmod \mathfrak{c}),$$

($a_{\mathfrak{c}}$ being the $\mathfrak{c}$-part of a) then for all $x \in G_{\mathbf{A}}$

$$\mathbf{f}(x\,w) = \psi(w^{\iota})\mathbf{f}(x) \quad \text{for } w \in W_{\mathfrak{c}} \text{ with } \ w_{\infty} = 1. \tag{1.4}$$

If $w = w(\theta) = (w_1(\theta_1), \cdots, w_n(\theta_n))$ where

$$w_{\nu}(\theta_{\nu}) = \begin{pmatrix} \cos\theta_{\nu} & -\sin\theta_{\nu} \\ \sin\theta_{\nu} & \cos\theta_{\nu} \end{pmatrix},$$

then

$$\mathbf{f}(x\,w(\theta)) = \mathbf{f}(x)e^{-ik\{\theta\}} \qquad (x \in G_{\mathbf{A}}). \tag{1.5}$$

An automorphic form $\mathbf{f}$ is called a *cusp form* if

$$\int_{F_{\mathbf{A}}/F} \mathbf{f}\left(\begin{pmatrix} 1 & t \\ 0 & 1 \end{pmatrix} g\right) dt = 0 \text{ for all } g \in G_{\mathbf{A}}. \tag{1.6}$$

The vector space $\mathcal{M}_k(\mathfrak{c}, \psi)$ of Hilbert automorphic forms of *holomorphic type* is defined as the set of functions satisfying (1.3) – (1.5) and the following *holomorphy condition* (1.7): for any $x \in G_{\mathbf{A}}$ with $x_{\infty} = 1$ there exists a holmorphic function $g_x : \mathfrak{H}^n \to \mathbf{C}$, such that for all $y = \left(\begin{smallmatrix} a & b \\ c & d \end{smallmatrix}\right) \in G_{\infty}^{+}$ we have

$$\mathbf{f}(x\,y) \quad = \quad (g_x|_k y)(\mathbf{i}) \tag{1.7}$$

(in the case $F = \mathbf{Q}$ we must also require that the functions g_x be holomorphic at the cusps). The property (1.7) enables one to describe the automorphic forms $\mathbf{f} \in \mathcal{M}_k(\mathfrak{c}, \psi)$ more explicitly in terms of Hilbert modular forms on $\mathfrak{H}^n$. For this purpose we put $f_{\lambda} = g_{x_{\lambda}^{-\iota}}$ where $x_{\lambda}^{-\iota} = \left(\begin{smallmatrix} t_{\lambda}^{-1} & 0 \\ 0 & 1 \end{smallmatrix}\right)$, then $f_{\lambda}(z) \in \mathcal{M}_k(\Gamma_{\lambda}, \psi_0)$ for the congruence subgroup

$$\begin{aligned} &\Gamma_{\lambda} = \Gamma_{\lambda}(\mathfrak{c}) \subset G_{\mathbf{Q}}^{+}, \\ &\Gamma_{\lambda} = x_{\lambda} W x_{\lambda}^{-1} \cap G_{\mathbf{Q}} = \\ &\left\{\begin{pmatrix} a & b \\ c & d \end{pmatrix} \in G_{\mathbf{Q}}^{+} | b \in \tilde{t}_{\lambda}^{-1}\mathfrak{d}^{-1},\ c \in \tilde{t}_{\lambda}\mathfrak{d}\mathfrak{c}, a, d \in \mathcal{O}_F,\ ad - bc \in \mathcal{O}_F^{\times}\right\}. \end{aligned}$$

This means that for all $\gamma \in \Gamma_\lambda(\mathfrak{c})$ the following condition (1.8) is satisfied:

$$f_\lambda|_k\gamma = \psi(\gamma)f_\lambda \quad \text{and} \quad f_\lambda(z) = \sum_\xi a_\lambda(\xi)e_F(\xi z), \tag{1.8}$$

where $0 \ll \xi \in \tilde{t}_\lambda$ or $\xi = 0$ in the sum over ξ (see [Shi6] for a more detailed discussion of Fourier expansions). The map $\mathbf{f} \mapsto (f_1, \cdots, f_h)$ defines a vector space isomorphism

$$\mathcal{M}(\mathfrak{c}, \psi) \cong \oplus_\lambda \mathcal{M}_k(\Gamma_\lambda, \psi)$$

Put

$$C(\mathfrak{m}, \mathbf{f}) = \begin{cases} a_\lambda(\xi)\mathcal{N}(\tilde{t}_\lambda)^{-k/2}, & \text{if the ideal } \mathfrak{m} = \xi\tilde{t}_\lambda^{-1} \text{ is integral;} \\ 0, & \text{if } \mathfrak{m} \text{ is not integral.} \end{cases} \tag{1.9}$$

We have the following *Fourier expansion*:

$$\mathbf{f}\left(\begin{pmatrix} y & x \\ 0 & 1 \end{pmatrix}\right) = \sum_{0 \ll \zeta \in F, \zeta = 0} C(\zeta\tilde{y}, \mathbf{f})|y|^{k/2} e_F(\zeta \mathbf{i} y_\infty)\chi(\zeta x), \tag{1.10}$$

where $\chi_F : F_\mathbf{A}/F \to \mathbf{C}^\times$ is a fixed additive character with the condition $\chi_F(x_\infty) = e_F(x_\infty)$ (see [Shi6], p. 650).

Let $\mathcal{S}_k(\mathfrak{c}, \psi) \subset \mathcal{M}(\mathfrak{c}, \psi)$ be the subspace of cusp forms and $\mathbf{f} \in \mathcal{S}_k(\mathfrak{c}, \psi)$ then $a_\lambda(0) = 0$ for all $\lambda = 1, \cdots, h$.

1.3. Hecke operators (see [Miy], [Shi1], [Shi6]) are introduced by means of double cosets of the type WyW for y in the semigroup

$$Y_\mathfrak{c} = G_\mathbf{A} \cap (G_\infty^+ \times \prod Y_\mathfrak{c}(\mathfrak{p})),$$

where

$$Y_\mathfrak{c}(\mathfrak{p}) = \left\{\begin{pmatrix} a & b \\ c & d \end{pmatrix} \in \mathrm{GL}_2(F_\mathfrak{p}) \mid a\mathcal{O}_\mathfrak{p} + \mathfrak{c}_\mathfrak{p} = \mathcal{O}_\mathfrak{p},\ b \in \mathfrak{d}_\mathfrak{p}^{-1}, c \in \mathfrak{c}_\mathfrak{p}\mathfrak{d}_\mathfrak{p}, d \in \mathcal{O}_\mathfrak{p}\right\}. \tag{1.11}$$

The Hecke algebra $\mathcal{H}_\mathfrak{c}$ consists of all formal finite sums of the type $\sum_y c_y WyW$ with $y \in Y_\mathfrak{c}, c_y \in \mathbf{C}$ and with the standard multiplication law defined by means of decomposition of double cosets into a disjoint union of a finite number of left cosets. By definition, $T_\mathfrak{c}(\mathfrak{m})$ is an element of the ring $\mathcal{H}_\mathfrak{c}$ obtained by taking the sum of all different WyW with $y \in Y_\mathfrak{c}$ such that $\widetilde{\det(y)} = \mathfrak{m}$. Let

$$T'_\mathfrak{c} = \mathcal{N}(\mathfrak{m})^{(k-2)/2}T_\mathfrak{c}(\mathfrak{m}) \tag{1.12}$$

be the normalized Hecke operator, whose action on the Fourier coefficients of an automorphic form (of the holomorphic type) $\mathbf{f} \in \mathcal{M}_k(\mathfrak{c}, \psi)$ is given by the usual formula

$$C(\mathfrak{m}, \mathbf{f}|T'_\mathfrak{c}(\mathfrak{m}) = \sum_{\mathfrak{m}+\mathfrak{n}=\mathfrak{a}} \psi^*(\mathfrak{a})\mathcal{N}(\mathfrak{a})^{k-1}C(\mathfrak{a}^{-2}\mathfrak{m}\mathfrak{n}, \mathbf{f}) \tag{1.13}$$

If $\mathbf{f} \in \mathcal{M}_k(\mathfrak{c}, \psi)$ is an eigenfunction of all Hecke operators $T'_{\mathfrak{c}}(\mathfrak{m})$ with $\mathbf{f}|_k T'_{\mathfrak{c}}(\mathfrak{m}) = \lambda(\mathfrak{m})\mathbf{f}$ then we have that $C(\mathfrak{m}, \mathbf{f}) = \lambda(\mathfrak{m}) C(\mathcal{O}_F, \mathbf{f})$. If we normalize the form $\mathbf{f}$ by the condition $C(\mathcal{O}_F, \mathbf{f}) = 1$ then the L-function has the following Euler product expansion:

$$L(s, \mathbf{f}) = \sum_{\mathfrak{n}} C(\mathfrak{n}, \mathbf{f}) \mathcal{N}(\mathfrak{n})^{-s} = \sum_{\mathfrak{n}} \lambda(\mathfrak{n}) \mathcal{N}(\mathfrak{n})^{-s} = \prod_{\mathfrak{p}} [1 - C(\mathfrak{p}, \mathbf{f}) \mathcal{N}(\mathfrak{p})^{-s} + \psi^*(\mathfrak{p}) \mathcal{N}(\mathfrak{p})^{k-1-2s}]^{-1}. \tag{1.14}$$

In this case the coefficients $C(\mathfrak{n}, \mathbf{f})$ of theform $\mathbf{f}$ are algebraic integers.

1.4. The Petersson inner product is defined for $\mathbf{f} = (f_1, \cdots, f_h) \in \mathcal{S}_k(\mathfrak{c}, \psi)$ and $\mathbf{g} = (g_1, \cdots, g_h) \in \mathcal{M}_k(\mathfrak{c}, \psi)$ by setting

$$\langle \mathbf{f}, \mathbf{g} \rangle_{\mathfrak{c}} = \sum_{\lambda=1}^{h} \int_{\Gamma_\lambda(\mathfrak{c}) \backslash \mathfrak{H}^n} \overline{f_\lambda(z)} g_\lambda(z) \mathcal{N}(y)^k \, d\mu(z), \tag{1.15}$$

where

$$d\mu(z) = \prod_{\nu=1}^{n} y_\nu^{-2} \, dx_\nu \, dy_\nu$$

is a G_∞^+ -invariant measure on $\mathfrak{H}^n$.

If $\mathbf{f} \in \mathcal{M}_k(\mathfrak{c}, \psi)$ and $\mathbf{f}|_k T'_{\mathfrak{c}}(\mathfrak{m}) = \lambda(\mathfrak{m})\mathbf{f}$ for all $\mathfrak{m}$ with $\mathfrak{m} + \mathfrak{c} = \mathbf{Q}_F$ then

$$\begin{gathered} \lambda(\mathfrak{m}) = \psi^*(\mathfrak{m}) \overline{\lambda(\mathfrak{m})}, \\ \psi^*(\mathfrak{m}) \langle \mathbf{f}|_k T'_\lambda(\mathfrak{m}), \mathbf{g} \rangle_{\mathfrak{c}} = \langle \mathbf{f}, \mathbf{g}|_k T'_\lambda(\mathfrak{m}) \rangle_{\mathfrak{c}} \end{gathered} \tag{1.16}$$

(ψ-hermitian property of the Hecke operators). Let $\mathfrak{q}$ be an integral ideal and $\mathbf{f} \in \mathcal{M}_k(\mathfrak{c}, \psi)$. Let us define the operators $\mathbf{f}|\mathfrak{q}$, $\mathbf{f}|U(\mathfrak{q})$ by their action on Fourier coefficients:

$$C(\mathfrak{m}, \mathbf{f}|\mathfrak{q}) = C(\mathfrak{q}^{-1}\mathfrak{m}, \mathbf{f}), \quad C(\mathfrak{m}, \mathbf{f}|U(\mathfrak{q})) = C(\mathfrak{q}\mathfrak{m}, \mathbf{f}). \tag{1.17}$$

Here is the explicit description of these operators: for a finite idele $q \in F_{\mathbf{A}}^\times$ with $\tilde{q} = \mathfrak{q}$

$$(\mathbf{f}|\mathfrak{q})(x) \quad = \quad \mathcal{N}(\mathfrak{q})^{-k/2} \mathbf{f}(x \begin{pmatrix} q & 0 \\ 0 & 1 \end{pmatrix}), \tag{1.18}$$

$$(\mathbf{f}|U(\mathfrak{q}))(x) = \mathcal{N}(\mathfrak{q})^{k/2-1} \sum_{v \in \mathcal{O}_F/\mathfrak{q}} \mathbf{f}(x \begin{pmatrix} 1 & v \\ 0 & q \end{pmatrix}). \tag{1.19}$$

We also define the involution $J_{\mathfrak{c}}$ by the formula

$$(\mathbf{f}|J_{\mathfrak{c}})(x) = \psi(\det(x)^{-1}) \mathbf{f}(x b_0) \text{ where } b_0 = \begin{pmatrix} 0 & 1 \\ c_0 & 0 \end{pmatrix} \in G_{\widehat{\mathbf{Q}}}, \tilde{c}_0 = \mathfrak{c}\mathfrak{d}^2, \tag{1.20}$$

then $\mathbf{f}|J_{\mathfrak{c}} \in \mathcal{M}_k(\mathfrak{c}, \psi^{-1})$. If f is a primitive form (in the sense of Miyake [Miy]) of conductor $\mathfrak{c}$ then the following hold

$$\mathbf{f}|J_{\mathfrak{c}} = \Lambda(\mathbf{f}) \mathbf{f}^\rho, \quad |\Lambda(\mathbf{f})| = 1, \; C(\mathfrak{m}, \mathbf{f}^\rho) = \overline{C(\mathfrak{m}, \mathbf{f})}. \tag{1.21}$$

From definitions (1.18)–(1.20) it follows that

$$\mathbf{f}|J_{\mathfrak{m}\mathfrak{c}} = \mathcal{N}(\mathfrak{m})^{k/2}(\mathbf{f}|J_{\mathfrak{c}})|\mathfrak{m}. \tag{1.22}$$

1.5. Gauss sums and the twist operator. Let χ be a Hecke character of finite order with a conductor $\mathfrak{m}$ and $\chi(x_\infty) = \mathrm{sign}(x_\infty)^r$ for $r = (r_1, \cdots, r_n) \in (\mathbf{Z}/2\mathbf{Z})^n$ (the parity of χ). Let χ^* be the character of the group of fractional ideals prime to $\mathfrak{m}$ which is associated to χ. Let us set $\chi^*(\mathfrak{a}) = 0$ for those $\mathfrak{a}$ which are not coprime to $\mathfrak{m}$ and define the Gauss sum by

$$\tau(\chi) = \sum_{x \in \mathfrak{m}^{-1}\mathfrak{d}^{-1}/\mathfrak{d}^{-1}} \mathrm{sign}(x)^r \chi^*((x)\mathfrak{m}\mathfrak{d}) e_F(x). \tag{1.23}$$

Then $|\tau(\chi)|^2 = \mathcal{N}(\mathfrak{m})$.

For any form $\mathbf{f} \in \mathcal{M}_k(\mathfrak{c}, \psi)$ there exists an automorphic form $\mathbf{f}(\chi) \in \mathcal{M}_k(\mathfrak{c}\mathfrak{m}^2, \psi\chi^2)$ which is uniquely defined by the condition $C(\mathfrak{n}, \mathbf{f}(\chi)) = \chi^*(\mathfrak{n})C(\mathfrak{n}, \mathbf{f})$ for all $\mathfrak{n}$ with $\mathfrak{n} + \mathfrak{m} = \mathcal{O}_F$. If a cusp form $\mathbf{f} \in \mathcal{S}_k(\mathfrak{c}, \psi)$ is a *primitive* of conductor $\mathfrak{c}$ and $\mathfrak{c} + \mathfrak{m} = \mathcal{O}_F$, then the conductor of $\mathbf{f}(\chi)$ is equal to $\mathfrak{c}\mathfrak{m}^2$ and the "pseudo-eigenvalue" of the involution $J_{\mathfrak{c}\mathfrak{m}^2}$ on $\mathbf{f}$ is given by the well known formula

$$\Lambda(\mathbf{f}(\chi)) = \psi^*(\mathfrak{m})\chi^*(\mathfrak{c})\tau(\chi)^2\mathcal{N}(\mathfrak{m})^{-1}\Lambda(\mathbf{f}) \tag{1.24}$$

(see [Shi6], p.664).

§2. Description of the non-Archimedean Rankin convolution of Hilbert automorphic forms.

2.1. As in the Introduction, let $\mathbf{f} \in \mathcal{S}_k(\mathfrak{c}(\mathbf{f}), \psi)$, $g \in \mathcal{S}_l(\mathfrak{c}(\mathbf{g}), \omega)$ be primitive cusp forms of (scalar) weights k and l, conductors $\mathfrak{c}(\mathbf{f})$, $\mathfrak{c}(\mathbf{g})$, and Hecke characters of finite order ψ and ω such that $k > l$ and $\psi(x) = \mathrm{sign}\mathcal{N}(x)^k$, $\omega(x) = \mathrm{sign}\mathcal{N}(x)^l$ for $x \in F_\infty^\times$. For an integral ideal $\mathfrak{a}$ denote by $S(\mathfrak{a})$ its support:

$$S(\mathfrak{a}) = \{\mathfrak{p} \mid \mathfrak{p} \text{ divides } \mathfrak{a}\},$$

and we set $S(\mathbf{f}) = S(\mathfrak{c}(\mathbf{f}))$, $S(\mathbf{g}) = S(\mathfrak{c}(\mathbf{g}))$ and $S(\chi) = S(\mathfrak{m})$ for a Hecke character of conductor $\mathfrak{m}$.

We shall make the following assumptions:

$$S_F \cap S(\mathbf{f}) = S_F \cap S(\mathbf{g}) = S(\mathbf{f}) \cap S(\mathbf{g}) = \emptyset, \tag{2.1}$$

$$C(\mathfrak{c}(\mathbf{f}), \mathbf{f}) \cdot C(\mathfrak{c}(\mathbf{g}), \mathbf{g}) \neq 0, \tag{2.2}$$

$$|i_p(C(\mathfrak{q}, \mathbf{f}))|_p = 1 \quad \text{for all } \mathfrak{q} \in S_F. \tag{2.3}$$

Put $\mathfrak{c} = \mathfrak{c}(\mathbf{f})\mathfrak{c}(\mathbf{g})$. For $\mathfrak{q} \in S_F$ we denote by the symbol $\alpha(\mathfrak{q})$ the root of the Hecke polynomial

$$X^2 - C(\mathfrak{q}, \mathbf{f})X + \psi^*(\mathfrak{q})\mathcal{N}(\mathfrak{q})^{k-1},$$

satisfying $|i_p(\alpha(\mathfrak{q}))|_p = 1$ and let $\alpha'(\mathfrak{q})$ be the other root of the polynomial. Then from the property (1.16) it follows that the numbers

$$\hat{\alpha}(\mathfrak{q}) = \overline{\psi^*(\mathfrak{q})}\alpha(\mathfrak{q}), \quad \hat{\alpha}'(\mathfrak{q}) = \overline{\psi^*(\mathfrak{q})}\alpha'(\mathfrak{q})$$

coincide with the roots of the complex conjugate polynomial

$$X^2 - \overline{C(\mathfrak{q},\mathbf{f})}X + \psi^*(\mathfrak{q})^{-1}\mathcal{N}(\mathfrak{q})^{k-1}$$

for all $\mathfrak{q}$ such that $\mathfrak{q} + \mathfrak{c}(\mathbf{f}) = \mathcal{O}_F$.

Similarly, if

$$X^2 - C(\mathfrak{q},\mathbf{g})X + \omega^*(\mathfrak{q})\mathcal{N}(\mathfrak{q})^{l-1} = (X - \beta(\mathfrak{q}))(X - \beta'(\mathfrak{q}))$$

then for $\mathfrak{q}$ with $\mathfrak{q} + \mathfrak{c}(\mathbf{g}) = \mathcal{O}_F$ the numbers

$$\hat{\beta}(\mathfrak{q}) = \overline{\omega^*(\mathfrak{q})}\beta(\mathfrak{q}), \quad \hat{\beta}'(\mathfrak{q}) = \overline{\omega^*(\mathfrak{q})}\beta'(\mathfrak{q})$$

coincide with the roots of the polynomial

$$X^2 - \overline{C(\mathfrak{q},\mathbf{g})}X + \omega^*(\mathfrak{q})^{-1}\mathcal{N}(\mathfrak{q})^{l-1}.$$

Let us extend the definition of the numbers $\alpha(\mathfrak{n}), \alpha'(\mathfrak{n}), \beta(\mathfrak{n}), \beta'(\mathfrak{n})$ to all integral ideals by multiplicativity. As in the introduction we use the same letter χ for both Hecke characters and the corresponding elements of $\mathcal{X}_S$, and set

$$\Omega(\mathbf{f}) = \langle \mathbf{f}, \mathbf{f} \rangle_{\mathfrak{c}(\mathbf{f})}.$$

2.2. Theorem. *Under assumptions (2.1) – (2.3) and notations as above there exists a bounded* $\mathbf{C}_p$*-analytic function* $\Psi_S : \mathcal{X}_S \to \mathbf{C}_p$ *which is uniquely determined by the condition: for all Hecke characters of finite order* $\chi \in \mathcal{X}_S^{\mathrm{tors}}$ *with* $S(\chi) \subset S_F$ *of conductor* $\mathfrak{m} = \mathfrak{c}(\chi)$ *and for all integers* $r = 0, 1, \cdots, k - l - 1$ *the value* $\Psi_S(\chi \mathcal{N} x_p^r)$ *is given by the image under the fixed embedding* i_p *of the algebraic number*

$$(-1)^{-r} D_F^{2r+2l} \omega^*(\mathfrak{m}) \frac{\tau(\chi)^2 \mathcal{N}\mathfrak{m}^{l+2r-1}}{\alpha(\mathfrak{m})^2} \frac{\Psi(l + r, \mathbf{f}, \mathbf{g}^\rho(\overline{\chi}))}{(-2\pi i)^{n(1-l)}\Omega(\mathbf{f})} A(r,\chi), \tag{2.4}$$

where

$$A(r,\chi) = \prod_{\mathfrak{q} \in S_F \setminus S(\chi)} (1 - (\chi^*\alpha^{-1}\beta)(\mathfrak{q})\mathcal{N}(\mathfrak{q})^r)(1 - (\chi^*\alpha^{-1}\beta')(\mathfrak{q})\mathcal{N}(\mathfrak{q})^r) \times \tag{2.5}$$
$$\times (1 - (\chi^{*-1}\alpha'\hat{\beta})(\mathfrak{q})\mathcal{N}(\mathfrak{q})^{-l-r})(1 - (\chi^{*-1}\alpha'\hat{\beta}')(\mathfrak{q})\mathcal{N}(\mathfrak{q})^{-l-r}),$$

D_F *is the discriminant of* F, $\tau(\chi)$ *is the Gauss sum of* χ, *and* $\mathbf{g}^\rho(\chi)$ *denotes the cusp form obtained from* $\mathbf{g}$ *by complex conjugation of its Fourier coefficients and by twisting with the character* χ (see §1).

§3. Distributions on the Galois group Gal_S.

3.1. The function $\Psi_S(x, \mathbf{f}, \mathbf{g})$ of theorem 2.2 is constructed using the theory of *non-Archimedean integration* ([Man6], [Ko1], [Ko2], [V1]). Let $\mathcal{O}_p$ be the subring of integral elements in the Tate field $\mathbf{C}_p$, $\mathcal{O}_p = \{x \in \mathbf{C}_p \mid |x|_p \le 1\}$ and $\mathcal{C}(\mathrm{Gal}_S, \mathbf{C}_p)$ denote the ring of all continuous p-adic functions on Gal_S. Then by a *$\mathcal{O}_p$-measure* on Gal_S we mean an arbitrary $\mathcal{O}_p$-linear map $\mu : \mathcal{C}(\mathrm{Gal}_S, \mathbf{C}_p) \to \mathcal{O}_p$ which is also denoted

$$\phi \quad \mapsto \quad \int_{\mathrm{Gal}_S} \phi \, d\mu.$$

We note that

$$\mathcal{X}_S = \mathrm{Hom}_{\mathrm{contin}}(\mathrm{Gal}_S, \mathbf{C}_p^\times) \subset \mathcal{C}(\mathrm{Gal}_S, \mathbf{C}_p).$$

Therefore for an arbitrary $\mathcal{O}_p$ -measure μ we can define the function

$$L_\mu : \mathcal{X}_S \to \mathcal{O}_p \quad (L_\mu = \int_{\mathrm{Gal}_S} \chi d\mu \quad \text{for} \quad \chi \in \mathcal{X}_S)$$

which is called *the non-Archimedean Mellin transform* of the measure μ. The function L_μ is bounded and analytic on the p-adic analytic Lie group $\mathcal{X}_S$. If we fix a character $\chi_0 \in \mathcal{X}_S$ then the set of values of the type $L_\mu(\chi_0\,\chi)$ for $\chi \in \mathcal{X}_S^{\mathrm{tors}}$ determines the function L_μ uniquely (see [V1], [Man6], [Ka3]). The analytic structure on $\mathcal{X}_S$ is given using class field theory. Namely, let $\mathfrak{m}$ be an integral ideal, $I(\mathfrak{m})$ be the group of fractional ideals coprime to $\mathfrak{m}$,

$$P(\mathfrak{m}) = \{(\alpha) \mid \alpha \in F_+^\times,\ \alpha \equiv 1(\mathrm{mod}^\times \mathfrak{m})\}$$
$$H(\mathfrak{m}) = I(\mathfrak{m})/\,P(\mathfrak{m})$$

the ideal class group (in the narrow sense) of conductor $\mathfrak{m}$. Then if $F(\mathfrak{m})/F$ is the maximal abelian extension of F unramified outside primes dividing $\mathfrak{m}$ and infinity then the Artin symbol provides an isomorphism:

$$H(\mathfrak{m}) \xrightarrow{\sim} \mathrm{Gal}(F(\mathfrak{m})/\,F), \quad \mathfrak{a} \mapsto \left(\frac{\mathfrak{a}, F(\mathfrak{m})}{F}\right),$$

and we have

$$F(S) = \cup_{\mathfrak{m}} F(\mathfrak{m}), \quad \mathrm{Gal}_S = \varprojlim_{\mathfrak{m}} H(\mathfrak{m}) \quad (S(\mathfrak{m}) \subset S_F).$$

Hence, there is the following exact sequence

$$1 \to \mathrm{Gal}_S^0 \to \mathrm{Gal}_S \to \mathrm{Gal}_1^+ \to 1$$

where $\mathrm{Gal}_1^+ = \mathrm{Gal}(F(\emptyset))/\,F)) = \widetilde{Cl}_F$, $F(\emptyset)$ being the maximal abelian extension of F ramified only over ∞,

$$\mathrm{Gal}_S^0 = \mathrm{Gal}(F(S)/F(\emptyset)) = \mathcal{O}_S^\times/(\mathrm{clos}(\mathcal{O}_+^\times)),$$

where $\mathrm{clos}(\mathcal{O}_+^\times)$ denotes the closure of the group of all totally positive units $\mathcal{O}_+^\times$ in the group

$$\mathcal{O}_S^\times = \prod_{q \in S} (\mathcal{O}_F \otimes \mathbf{Z}_q)^\times.$$

There is a canonical $\mathbf{C}_p$-analytic Lie group structure on the group

$$\mathrm{Hom}_{\mathrm{contin}}((\mathcal{O}_F \otimes \mathbf{Z}_p)^\times, \mathbf{C}_p^\times).$$

We extend this structure to the entire group $\mathcal{X}_S$ using shifts (see [Man6], [V1]).

3.2. We will construct a measure $\mu = \mu(\mathbf{f}, \mathbf{g})$ on the group Gal_S for which the value $L_\mu(\chi \mathcal{N} x_p^r)$ is given by the formula (2.4) and we set $\Psi_S(x) = L_\mu(x)$. In order to construct μ we use the theory of *distributions*. Recall that for an abelian group $\mathcal{A}$ the *$\mathcal{A}$-distribution* μ on Gal_S is a finite additive system of functions

$$\mu = \{\mu_{\mathfrak{m}}\}_\mu, \quad \mu_{\mathfrak{m}} : H(\mathfrak{m}) \to \mathcal{A}$$

satisfying the following compatibility condition:

$$\sum_{x \equiv y \bmod{}^\times \mathfrak{m}_1} \mu_{\mathfrak{m}}(x) = \mu_{\mathfrak{m}_1}(y) \tag{3.1}$$

for all $\mathfrak{m}_1$ with $\mathfrak{m}_1 | \mathfrak{m}$. If $\mathcal{A} = \mathcal{O}_p$ then μ determines a $\mathcal{O}_p$-measure. We call a $\mathbf{C}_p$-distribution a *bounded measure* if $a\mu$ is a $\mathcal{O}_p$ -measure for some $a \in \mathbf{C}_p^\times$. Now let $\mathcal{A}$ be a vector space over $\mathbf{Q}^{\mathrm{ab}}$ (the maximal abelian extension of $\mathbf{Q}$) and $\mu_{\mathfrak{m}} : H(\mathfrak{m}) \to \mathcal{A}$ be an arbitrary system of functions. For a character $\chi : H(\mathfrak{m}_1) \to \mathbf{Q}^{\mathrm{ab}\times}$ and for $\mathfrak{m}_1 | \mathfrak{m}$ we define $\chi_{\mathfrak{m}}$ to be composition of the natural projection $H(\mathfrak{m}) \to H(\mathfrak{m}_1)$ $(\bmod \mathfrak{m}_1)$ and χ and put

$$\mu_{\mathfrak{m}}(\chi_{\mathfrak{m}}) = \sum_{x \in H(\mathfrak{m})} \chi_{\mathfrak{m}}(x) \mu_{\mathfrak{m}}(x).$$

Then the compatibility condition (3.1) is equivalent to the fact that

$$\mu_{\mathfrak{m}}(\chi_{\mathfrak{m}}) \text{ does not depend on the choice of } \mathfrak{m} \text{ with } \mathfrak{m}_1 | \mathfrak{m} \tag{3.2}$$

(see also Proposition 2.2 of chapter 3).

Using this compatibility criterion we construct $\mu(\mathbf{f}, \mathbf{g})$ by first constructing a family of distributions

$$\tilde{\mu}_{s,\mathfrak{m}} : H(\mathfrak{m}) \to \overline{\mathbf{Q}},$$

which is parametrized by the complex parameter s.

3.3. Proposition-Definition. *For every $s \in \mathbf{C}$ there exists a complex valued distribution $\{\tilde{\mu}_{s,\mathfrak{m}}\}_{\mathfrak{m}}$ which is uniquely determined by the condition*

$$\tilde{\mu}_{s,\mathfrak{m}}(\chi_{\mathfrak{m}}) = \frac{\mathcal{N}(\mathfrak{c}\mathfrak{d}^2)^s \mathcal{N}(\mathfrak{m}')^{s-(l/2)}}{\alpha(\mathfrak{m}')} \frac{\Psi(s, \mathbf{f}_0 | \mathfrak{c}(\mathbf{g}), \mathbf{g}(\chi_{\mathfrak{m}}) | J_{\mathfrak{c}\mathfrak{m}'})}{(-2\pi i)^{n(1-l)} \cdot \Omega(\mathbf{f})} \tag{3.3}$$

where $\mathfrak{m}, \mathfrak{m}'$ are arbitrary ideals with the condition $\mathfrak{m}_0 \mathfrak{c}(\chi) | \mathfrak{m}$, $\mathfrak{c}(\chi)^2 \mathfrak{m}_0^2 | \mathfrak{m}'$,

$$\mathbf{f}_0 = \sum_{\mathfrak{a} | \mathfrak{m}_0} \mathrm{M}(\mathfrak{a}) \alpha'(\mathfrak{a})\, \mathbf{f} | \mathfrak{a},$$

with $\mathrm{M}(\mathfrak{a})$ being the Möbius function (of ideals) so that this definition of $\mathbf{f}_0$ is equivalent to the identity

$$L(s, \mathbf{f}_0) = \prod_{\mathfrak{q} | \mathfrak{m}_0} (1 - \alpha'(\mathfrak{q}) \mathcal{N}(\mathfrak{q})^{-s}) L(s, \mathbf{f}).$$

Note that in the Hilbert modular case the fact that the function $\mathbf{f}_0$ does not vanish follows from the explicit description of its Fourier coefficients (so that they essentially coincide with the eigenvalues of the corresponding Hecke operators).

3.4. *The proof* of the proposition is carried out by means of the compatibility criterion (3.2) according to which it suffices to verify that the right hand side of (3.3) does not depend on $\mathfrak{m}$ and on $\mathfrak{m}'$. First of all we note that $\mathbf{g}(\chi_{\mathfrak{m}}) = \mathbf{g}(\chi_{\mathfrak{m}_0})$ does not depend on $\mathfrak{m}$ since $S(\mathfrak{m}_0) = S(\mathfrak{m}) = S_F$ and

$$L(s, \mathbf{g}(\chi_{\mathfrak{m}})) = \sum_{\mathfrak{n}+\mathfrak{m}_0=\mathcal{O}_F} \chi(\mathfrak{n})C(\mathfrak{n}, \mathbf{g})\mathcal{N}(\mathfrak{n})^{-s}.$$

On the other hand, the formula (1.22) implies the identity: for $\mathfrak{m}' = \mathfrak{m}_0^2\mathfrak{c}(\chi)^2\mathfrak{m}_1$ we have that

$$\mathbf{g}(\chi_{\mathfrak{m}})|J_{\mathfrak{c}\mathfrak{m}'} = \mathcal{N}(\mathfrak{m}_1\mathfrak{c}(\mathbf{f}))^{l/2}\,(\mathbf{g}_{\mathfrak{m}_0}(\chi)|J_{\mathfrak{c}(\mathbf{g})\mathfrak{m}_0^2\mathfrak{c}(\chi)^2})|\mathfrak{m}_1\mathfrak{c}(\mathbf{f}). \tag{3.4}$$

If we set

$$\sum_{\mathfrak{n}} A(\mathfrak{n})\mathcal{N}(\mathfrak{n})^{-s} = L(s, \mathbf{f}_0), \tag{3.5}$$

$$\sum_{\mathfrak{n}} B(\mathfrak{n})\mathcal{N}(\mathfrak{n})^{-s} = L(s, \mathbf{g}_{\mathfrak{m}_0}(\chi)|J_{\mathfrak{c}(\mathbf{g})\mathfrak{m}_0^2\mathfrak{c}(\chi)^2}), \tag{3.6}$$

then (1.17) and (1.21) imply the equalities

$$A(\mathfrak{n}\mathfrak{c}(\mathbf{f})\mathfrak{m}_1) = \alpha(\mathfrak{c}(\mathbf{f})\mathfrak{m}_1)A(\mathfrak{n}) = C(\mathfrak{c}(\mathbf{f}), \mathbf{f})\alpha(\mathfrak{m}_1)A(\mathfrak{n}), \tag{3.7}$$

$$B(\mathfrak{n}\mathfrak{c}(\mathbf{g})) = \overline{C(\mathfrak{c}(\mathbf{g}), \mathbf{g})}\chi^{*-1}(\mathfrak{c}(\mathbf{g}))B(\mathfrak{n}). \tag{3.8}$$

Taking into account that $\mathfrak{m}_1\mathfrak{c}(\mathbf{f}) + \mathfrak{c}(\mathbf{g}) = \mathcal{O}_F$ we obtain the following equality from (3.4), (3.7), and(3.8):

$$\begin{aligned} &L(s, \mathbf{f}_0|\mathfrak{c}(\mathbf{g}), g(\chi_{\mathfrak{m}_0})|J_{\mathfrak{c}\mathfrak{m}'}) = \\ &\mathcal{N}(\mathfrak{m}_1\mathfrak{c}(\mathbf{f}))^{l/2}\sum_{\mathfrak{n}} A(\mathfrak{n}\mathfrak{c}(\mathbf{g})^{-1})B(\mathfrak{n}\mathfrak{m}_1^{-1}\mathfrak{c}(\mathbf{f})^{-1})\mathcal{N}(\mathfrak{n})^{-s} = \\ &\mathcal{N}(\mathfrak{m}_1\mathfrak{c}(\mathbf{f}))^{l/2}\sum_{\mathfrak{n}} A(\mathfrak{n}\mathfrak{m}_1\mathfrak{c}(\mathbf{f}))B(\mathfrak{n}\mathfrak{m}_1\mathfrak{c}(\mathbf{g}))\mathcal{N}(\mathfrak{n}\mathfrak{m}_1\mathfrak{c}(\mathbf{f})\mathfrak{c}(\mathbf{g}))^{-s} = \\ &\frac{\kappa(\mathbf{f}, \mathbf{g})\alpha(\mathfrak{m}_1)\chi^{*-1}(\mathfrak{c}(\mathbf{g}))}{\mathcal{N}(\mathfrak{c})^s\mathcal{N}(\mathfrak{m}_1)^{s-(l/2)}} L(s, \mathbf{f}_0, g(\chi_{\mathfrak{m}_0})|J_{\mathfrak{c}(\mathbf{g})\mathfrak{c}(\chi)^2\mathfrak{m}_0^2}) \end{aligned} \tag{3.9}$$

with a constant

$$\kappa(\mathbf{f}, \mathbf{g}) = \mathcal{N}(\mathfrak{c}(\mathbf{f}))^{l/2}C(\mathfrak{c}(\mathbf{f}), \mathbf{f})\overline{C(\mathfrak{c}(\mathbf{g}), \mathbf{g})}$$

which is not zero due to the assumption (2.2). From (3.9) and from definition (0.2) it follows now that

$$\begin{aligned} &\frac{\mathcal{N}(\mathfrak{c}\mathfrak{d}^2)^s\mathcal{N}(\mathfrak{m}')^{s-(l/2)}}{\alpha(\mathfrak{m}')}\Psi(s, \mathbf{f}_0|\mathfrak{c}(\mathbf{g}), g(\chi_{\mathfrak{m}})|J_{\mathfrak{c}\mathfrak{m}'}) = \\ &\kappa(\mathbf{f}, \mathbf{g})D_F^{2s}\frac{\mathcal{N}(\mathfrak{c}(\chi)\mathfrak{m}_0)^{2s-l}\chi^*(\mathfrak{c}(\mathbf{g}))^{-1}}{\alpha(\mathfrak{c}(\chi)\mathfrak{m}_0)^2}\cdot\Psi(s, \mathbf{f}_0, \mathbf{g}(\chi_{\mathfrak{m}_0})|J_{\mathfrak{c}(\mathbf{g})\mathfrak{c}(\chi)^2\mathfrak{m}_0^2}) \end{aligned} \tag{3.10}$$

It is easy to notice that the terms containing $\mathfrak{m}'$ in (3.10) cancel. Hence, $\tilde{\mu}_{s,\mathfrak{m}}(\chi_{\mathfrak{m}})$ does not depend on $\mathfrak{m}$ or $\mathfrak{m}'$ proving the proposition.

Next we will use the property (1.24) of the twist operator in order to obtain a more explicit expression for the right hand side of (3.3), which is essential for establishing the precise formula (2.4) in theorem 2.2.

3.5. Proposition. *Under the notations and assumptions as above the following equality holds*

$$\Psi(s, \mathbf{f}_0, \mathbf{g}(\chi_{\mathfrak{m}_0}) | J_{\mathfrak{c}(\mathbf{g})\mathfrak{c}(\chi)^2\mathfrak{m}_0^2}) = \alpha(\mathfrak{m}_0)^2 \mathcal{N}(\mathfrak{m}_0)^{l-2s} A(s-l,\chi)\Lambda(\mathbf{g}(\chi))\Psi(s, \mathbf{f}, \mathbf{g}^\rho(\chi^{-1})), \tag{3.11}$$

where

$$A(s-l,\chi) = \prod_{\mathfrak{q}\in S_F \setminus S(\chi)} (1-(\chi^*\alpha^{-1}\beta)(\mathfrak{q})\mathcal{N}(\mathfrak{q})^{s-l})(1-(\chi^*\alpha^{-1}\beta')(\mathfrak{q})\mathcal{N}(\mathfrak{q})^{s-l}) \times \times (1-(\chi^{*-1}\alpha'\hat{\beta})(\mathfrak{q})\mathcal{N}(\mathfrak{q})^{-s})(1-(\chi^{*-1}\alpha'\hat{\beta}')(\mathfrak{q})\mathcal{N}(\mathfrak{q})^{-s}) \tag{3.12}$$

is defined by (2.5) and the twisted root number

$$\Lambda(\mathbf{g}(\chi)) = \omega^*(\mathfrak{c}(\chi))\chi^*(\mathfrak{c}(\mathbf{g}))\tau(\chi)^2\mathcal{N}(\mathfrak{c}(\chi))^{-1}\Lambda(\mathbf{g})$$

is given by (1.24), so that

$$\mathbf{g}(\chi)|J_{\mathfrak{c}(\mathbf{g})\mathfrak{c}(\chi)^2} = \Lambda(\mathbf{g}(\chi))\mathbf{g}^\rho(\overline{\chi}).$$

3.6. In order to prove the proposition we use the Möbius function $M(\mathfrak{a})$ (of ideals $\mathfrak{a} \subset \mathcal{O}_F$). By definition,

$$\sum_{\mathfrak{n}} M(\mathfrak{n})\mathcal{N}(\mathfrak{n})^{-s} = \prod_{\mathfrak{p}} (1-\mathcal{N}\mathfrak{p}^{-s}),$$

which is equivalent to saying that that

$$\sum_{\mathfrak{n}_1|\mathfrak{n}} M(\mathfrak{n}_1) = \begin{cases} 1, & \text{if } \mathfrak{n} = \mathcal{O}_F; \\ 0, & \text{otherwise.} \end{cases} \tag{3.13}$$

We apply this property to the Fourier expansion of $\mathbf{g}(\chi_{\mathfrak{m}_0})$ which is given for all $\lambda = 1, \cdots, h$ by

$$\mathbf{g}(\chi_{\mathfrak{m}_0})_\lambda(z) = \sum_{\xi} b_\lambda(\xi)\chi^*(\xi\tilde{t}_\lambda^{-1})e_F(\xi z),$$

where the sum is extended over all $\xi \in \tilde{t}_\lambda$ such that $\xi \gg 0$, $\tilde{\xi} + \mathfrak{m}_0 = \mathcal{O}_F$, and

$$b_\lambda(\xi) = \mathcal{N}(\tilde{t}_\lambda)^{l/2}C(\xi\tilde{t}_\lambda^{-1}, \mathbf{g})$$

are the Fourier coefficients (see (1.8) - (1.10)). Using (3.13) we see that

$$\mathfrak{g}(\chi_{\mathfrak{m}_0}) = \sum_{\mathfrak{n}_1|\mathfrak{m}_0} M(\mathfrak{n}_1)\mathfrak{g}(\chi)|U(\mathfrak{n}_1)\cdot\mathfrak{n}_1, \tag{3.14}$$

since

$$C(\mathfrak{n}, \mathfrak{g}(\chi_{\mathfrak{m}_0})) = \sum_{\mathfrak{n}_1|(\mathfrak{m}_0+\mathfrak{n})} M(\mathfrak{n}_1)C(\mathfrak{n}, \mathfrak{g}(\chi)).$$

Let us now express the operators $U(\mathfrak{n}_1)$ in terms of the Hecke operators: for all prime ideals $\mathfrak{q}$ not dividing $\mathfrak{c}(\mathfrak{g})$ we have that

$$\begin{aligned} &\mathfrak{g}(\chi)|T'(\mathfrak{q}) = \mathfrak{g}(\chi)|U(\mathfrak{q}) + (\omega^*\chi^{2*})(\mathfrak{q})\mathcal{N}\mathfrak{q}^{l-1}\mathfrak{g}(\chi)|\mathfrak{q} = \chi^*(\mathfrak{q})C(\mathfrak{q},\mathfrak{g})\mathfrak{g}(\chi), \\ &\mathfrak{g}(\chi)|U(\mathfrak{n}_1) = \mathfrak{g}(\chi)|\prod_{\mathfrak{q}|\mathfrak{n}_1}(\chi^*(\mathfrak{q})C(\mathfrak{q},\mathfrak{g}) - (\omega^*\chi^{*2})(\mathfrak{q})\mathcal{N}\mathfrak{q}^{l-1}\cdot\mathfrak{q}) = \\ &\mathfrak{g}(\chi)|\left(\sum_{\mathfrak{n}_2|\mathfrak{n}_1} M(\mathfrak{n}_2)\omega^*(\mathfrak{n}_2)\mathcal{N}n_2^{l-1}C(\mathfrak{n}_1\mathfrak{n}_2^{-1},\mathfrak{g})\chi^*(\mathfrak{n}_1\mathfrak{n}_2)\cdot\mathfrak{n}_2\right), \end{aligned} \tag{3.15}$$

since $\mathfrak{n}_1$ is square free and $\mathfrak{n}_1 + \mathfrak{c}(\mathfrak{g}) = \mathcal{O}_F$. From (3.14) and (3.15) it now follows that

$$\mathfrak{g}(\chi_{\mathfrak{m}_0}) = \sum_{\mathfrak{n}_2|\mathfrak{n}_1|\mathfrak{m}_0} M(\mathfrak{n}_2)M(\mathfrak{n}_1)\omega^*(\mathfrak{n}_2)\mathcal{N}n_2^{l-1}C(\mathfrak{n}_1\mathfrak{n}_2^{-1},\mathfrak{g})\chi^*(\mathfrak{n}_1\mathfrak{n}_2)\mathfrak{g}(\chi)|(\mathfrak{n}_1\mathfrak{n}_2). \tag{3.16}$$

Let us apply the involution $J_{\mathfrak{c}(\mathfrak{g})\mathfrak{c}(\chi)^2}$ to the function (3.16) using the following apparent commutation relation of involutions and shifts, which is analogous to (1.22): for all integral ideals $\mathfrak{a}$, $\mathfrak{a}'$ and $\mathfrak{b}$ such that $\mathfrak{a}'|\mathfrak{b}$ we have that

$$\mathfrak{g}|\mathfrak{a}'\cdot J_{\mathfrak{a}\mathfrak{b}} = \mathcal{N}\mathfrak{a}^{l/2}\mathcal{N}\mathfrak{a}'^{-l}\mathfrak{g}|J_{\mathfrak{b}}\cdot(\mathfrak{a}\mathfrak{a}'^{-1}). \tag{3.17}$$

If we now substitute in (3.17) $\mathfrak{a} = \mathfrak{m}_0$, $\mathfrak{a}' = \mathfrak{n}_1\mathfrak{n}_2$, $\mathfrak{b} = \mathfrak{c}(\mathfrak{g})\mathfrak{c}(\chi)^2$, then we get the following relation

$$\begin{aligned} \mathfrak{g}(\chi)|(\mathfrak{n}_1\mathfrak{n}_2)J_{\mathfrak{c}(\mathfrak{g})\mathfrak{m}_0^2\mathfrak{c}(\chi)^2} =& \mathcal{N}\mathfrak{m}_0^l\mathcal{N}(\mathfrak{n}_1\mathfrak{n}_2)^{-l}\mathfrak{g}(\chi)|J_{\mathfrak{c}(\mathfrak{g})\mathfrak{c}(\chi)^2)}(\mathfrak{m}_0^2\mathfrak{n}_1^{-1}\mathfrak{n}_2^{-1}) = \\ & \mathcal{N}\mathfrak{m}_0^l\mathcal{N}(\mathfrak{n}_1\mathfrak{n}_2)^{-l}\Lambda(\mathfrak{g}(\chi))\mathfrak{g}^{\rho}(\overline{\chi})|(\mathfrak{m}_0^2\mathfrak{n}_1^{-1}\mathfrak{n}_2^{-1}). \end{aligned} \tag{3.18}$$

Substituting (3.18) into (3.16) gives

$$\begin{aligned} &\mathfrak{g}(\chi_{\mathfrak{m}_0})|J_{\mathfrak{c}(\mathfrak{g})\mathfrak{m}_0^2\mathfrak{c}(\chi)^2} = \\ &\quad \mathcal{N}\mathfrak{m}_0^l\Lambda(\mathfrak{g}(\chi))\sum_{\mathfrak{n}_2|\mathfrak{n}_1|\mathfrak{m}_0} M(\mathfrak{n}_2)M(\mathfrak{n}_1)\omega^*(\mathfrak{n}_2)\mathcal{N}n_2^{l-1}C(\mathfrak{n}_1\mathfrak{n}_2^{-1},\mathfrak{g})\times \\ &\quad \times\chi^*(\mathfrak{n}_1\mathfrak{n}_2)\mathcal{N}(\mathfrak{n}_1\mathfrak{n}_2)^{-l}\mathfrak{g}^{\rho}(\overline{\chi})|(\mathfrak{m}_0^2\mathfrak{n}_1^{-1}\mathfrak{n}_2^{-1}). \end{aligned} \tag{3.19}$$

We now apply (3.19) to (3.12) and notice first that

$$L(s,\mathbf{f}_0,\mathbf{g}(\chi_{\mathfrak{m}_0})|J_{\mathfrak{c}(\mathbf{g})\mathfrak{c}(\chi)^2\mathfrak{m}_0^2}) =$$
$$\mathcal{N}\mathfrak{m}_0^l\Lambda(\mathbf{g}(\chi))\sum_{\mathfrak{n}_2|\mathfrak{n}_1|\mathfrak{m}_0} M(\mathfrak{n}_2)M(\mathfrak{n}_1)\omega^*(\mathfrak{n}_2)\mathcal{N}n_2^{l-1}C(\mathfrak{n}_1\mathfrak{n}_2^{-1},\mathbf{g})\times \tag{3.20}$$
$$\times\,\chi^*(\mathfrak{n}_1\mathfrak{n}_2)\mathcal{N}(\mathfrak{n}_1\mathfrak{n}_2)^{-l}L(s,\mathbf{f}_0,\mathbf{g}^\rho(\overline{\chi})|(\mathfrak{m}_0^2\mathfrak{n}_1^{-1}\mathfrak{n}_2^{-1})).$$

If we consider coefficients of the series (3.20) we deduce directly from their definition that

$$L(s,\mathbf{f}_0,\mathbf{g}^\rho(\overline{\chi})|(\mathfrak{m}_0^2\mathfrak{n}_1^{-1}\mathfrak{n}_2^{-1})) = (\mathcal{N}\mathfrak{m}_0^2\mathfrak{n}_1^{-1}\mathfrak{n}_2^{-1})^{-s}L(s,\mathbf{f}_0|U(\mathfrak{m}_0^2\mathfrak{n}_1^{-1}\mathfrak{n}_2^{-1}),\mathbf{g}^\rho(\overline{\chi})) =$$

$$\alpha(\mathfrak{m}_0^2\mathfrak{n}_1^{-1}\mathfrak{n}_2^{-1})(\mathcal{N}\mathfrak{m}_0^2\mathfrak{n}_1^{-1}\mathfrak{n}_2^{-1})^{-s}L(s,\mathbf{f}_0,\mathbf{g}^\rho(\overline{\chi})). \tag{3.21}$$

Therefore the convolution (3.12) becomes

$$\Psi(s,\mathbf{f}_0,\mathbf{g}(\chi_{\mathfrak{m}_0})|J_{\mathfrak{c}(\mathbf{g})\mathfrak{c}(\chi)^2\mathfrak{m}_0^2}) = \mathcal{N}\mathfrak{m}_0^{l-2s}\alpha(\mathfrak{m}_0)^2\Lambda(\mathbf{g}(\chi))\Psi(s,\mathbf{f}_0,\mathbf{g}^\rho(\overline{\chi}))\times$$
$$\sum_{\mathfrak{n}_2|\mathfrak{n}_1|\mathfrak{m}_0} M(\mathfrak{n}_2)M(\mathfrak{n}_1)\omega^*(\mathfrak{n}_2)\mathcal{N}n_2^{l-1}\chi^*(\mathfrak{n}_1\mathfrak{n}_2)\alpha(\mathfrak{n}_1\mathfrak{n}_2)^{-1}(\mathcal{N}\mathfrak{n}_1\mathfrak{n}_2)^{s-l}C(\mathfrak{n}_1\mathfrak{n}_2^{-1},\mathbf{g}) \tag{3.22}$$

The series (3.22) is an Euler product in which Euler factors can be easily written using (0.6) and the definition of $\mathbf{f}_0$ (in proposition 3.3):

$$\Psi(s,\mathbf{f}_0,\mathbf{g}^\rho(\overline{\chi})) = \Psi(s,\mathbf{f},\mathbf{g}^\rho(\overline{\chi}))\times$$
$$\times\prod_{\mathfrak{q}\in S_F\setminus S(\chi)}\Big((1-(\chi^{*-1}\alpha'\hat{\beta})(\mathfrak{q})\mathcal{N}(\mathfrak{q})^{-s})(1-(\chi^{*-1}\alpha'\hat{\beta}')(\mathfrak{q})\mathcal{N}(\mathfrak{q})^{-s})\Big). \tag{3.23}$$

Next we notice that by definition of the Hecke polynomial we have

$$X^2 - C(\mathfrak{q},\mathbf{g})X + \omega^*(\mathfrak{q})\mathcal{N}(\mathfrak{q})^{l-1} = (X-\beta(\mathfrak{q}))(X-\beta'(\mathfrak{q})),$$

which implies the equality

$$\sum_{\mathfrak{n}_2|\mathfrak{n}_1|\mathfrak{m}_0} M(\mathfrak{n}_2)M(\mathfrak{n}_1)\omega^*(\mathfrak{n}_2)\mathcal{N}n_2^{l-1}C(\mathfrak{n}_1\mathfrak{n}_2^{-1},\mathbf{g})\chi^*(\mathfrak{n}_1\mathfrak{n}_2)\alpha(\mathfrak{n}_1\mathfrak{n}_2)^{-1}(\mathfrak{n}_1\mathfrak{n}_2)^{s-l} =$$
$$\prod_{\mathfrak{q}\in S_F\setminus S(\chi)}(1-(\chi^*\alpha^{-1}\beta)(\mathfrak{q})\mathcal{N}(\mathfrak{q})^{s-l})(1-(\chi^*\alpha^{-1}\beta')(\mathfrak{q})\mathcal{N}(\mathfrak{q})^{s-l}). \tag{3.24}$$

The proposition follows, if we combine the equalities (3.22) -(3.24).

3.7. Let us now substitute (3.11) in the definition (3.3):

$$\tilde{\mu}_{s,\mathfrak{m}}(\chi_\mathfrak{m}) = \kappa(\mathbf{f},\mathbf{g})\Lambda(\mathbf{g})A(s-l,\chi)D_F^{2s}\times$$
$$\times\,\frac{\omega^*(\mathfrak{c}(\chi))\tau(\chi)^2\mathcal{N}(\mathfrak{c}(\chi))^{2s-l-1}}{\alpha(\mathfrak{c}(\chi))^2}\cdot\frac{\Psi(s,\mathbf{f},\mathbf{g}^\rho(\chi^{-1}))}{(-2\pi i)^{n(1-l)}\Omega(\mathbf{f})}. \tag{3.25}$$

Now we see from the algebraicity property (0.3) that the values $\tilde{\mu}_{l+r,\mathfrak{m}}(y)$ belong to $\overline{\mathbf{Q}}$ for all $y \in H(\mathfrak{m})$. Setting

$$\mu(y, \mathbf{f}, \mathbf{g}) = i_p(\tilde{\mu}_l(y)/\Lambda(\mathbf{g})\kappa(\mathbf{f}, \mathbf{g})), \tag{3.26}$$

as a result we obtain a $\mathbf{C}_p$-valued distribution on the Galois group Gal_S which satisfies the equality (2.4) with $r = 0$. The p-adic boundedness of the distribution is equivalent to certain generalized Kummer congruences for the numbers

$$\frac{\Psi(s, \mathbf{f}, \mathbf{g}^\rho(\chi^{-1}))}{(-2\pi i)^{n(1-l)}\Omega(\mathbf{f})},$$

and will be established in §5 below.

§4. The integral representation of Rankin – Shimura and the holomorphic projection operator.

4.1. We start by recalling the definition of *Eisenstein series* in the Hilbert modular case. Let $\mathfrak{a}, \mathfrak{b}$ be arbitrary fractional ideals, m a positive integer, $q \in (q_1, \cdots, q_n) \in \mathbf{Z}^n$, $q_\nu \geq 0$, η a Hecke character of finite order modulo an integral ideal $\mathfrak{c} \subset \mathcal{O}_F$ such that $\eta^*((x)) = \mathrm{sign}\mathcal{N}(x)^m$ for $x \equiv 1 \bmod {}^\times \mathfrak{c}$, $x \in \mathcal{O}_F$. We put (for $\mathrm{Re}(s) > 2 - m$)

$$\begin{aligned} K^q_m(z, s; \mathfrak{a}, \mathfrak{b}; \eta) &= (2\pi i)^{-\{q\}}(z - \overline{z})^{-q} \times \\ &\times \sum_{c,d} \mathrm{sign}\mathcal{N}(d)^m\, \eta^*(d\mathfrak{b}^{-1}) \left(\frac{c\overline{z} + d}{cz + d}\right)^q \mathcal{N}(cz + d)^{-m}|\mathcal{N}(cz + d)|^{-2s}, \end{aligned} \tag{4.1}$$

$$\begin{aligned} L^q_m(z, s; \mathfrak{a}, \mathfrak{b}; \eta) &= (2\pi i)^{-\{q\}}(z - \overline{z})^{-q} \times \\ &\times \sum_{c,d} \mathrm{sign}\mathcal{N}(c)^m\, \eta^*(c\mathfrak{a}^{-1}) \left(\frac{c\overline{z} + d}{cz + d}\right)^q \mathcal{N}(cz + d)^{-m}|\mathcal{N}(cz + d)|^{-2s}, \end{aligned} \tag{4.2}$$

where $z^q = \prod_\nu z_\nu^{q_\nu}$, $\mathcal{N}(z) = z_1 \cdots z_n$, the summation in (4.1) and (4.2) is taken over a system of representatives (c, d) of $\mathcal{O}_F^\times$-equivalence classes of non-zero elements in $\mathfrak{a} \times \mathfrak{b}$ ($(c, d) \sim (uc, ud)$ for $u \in \mathcal{O}_F^\times$). The series (4.1) and (4.2) can be extended to functions on the adelic group $G_\mathbf{A}$ as in §1 in such a way that

$$K^q_m(s; \mathfrak{a}, \mathfrak{b}; \eta)_\lambda(z) = \mathcal{N}(\tilde{t}_\lambda)^{s+(m/2)}\mathcal{N}(y)^s K^q_m(z, s; \tilde{t}_\lambda \mathfrak{d}\mathfrak{a}, \mathfrak{b}; \eta), \tag{4.3}$$

$$L^q_m(s; \mathfrak{a}, \mathfrak{b}; \eta)_\lambda(z) = \mathcal{N}(\tilde{t}_\lambda)^{-s-(m/2)}\mathcal{N}(y)^s L^q_m(z, s; \mathfrak{a}, \mathfrak{b}\tilde{t}_\lambda^{-1}\mathfrak{d}^{-1}; \eta). \tag{4.4}$$

The functions (4.1)–(4.4) admit analytic continuation onto the entire complex plane with respect to the parameter $s \in \mathbf{C}$, and they satisfy the following functional equation (see [Shi6], p.672, where it is given in a slightly different notation) under the assumption that η is *primitive* modulo $\mathfrak{c}$:

$$\begin{aligned} \Delta^q_m(z, 1 - m - s)K^q_m(1 - m - s; \mathfrak{a}, \mathfrak{b}; \eta) &= \\ \tau(\eta)\mathcal{N}(\mathfrak{d}\mathfrak{a}\mathfrak{b}\mathfrak{c})^{m+2s-1}&\Delta_m(s)^q L^q_m(s; \mathfrak{a}, \mathfrak{b}\mathfrak{c}; \overline{\eta}), \end{aligned} \tag{4.5}$$

with the Γ-factor $\Delta^q_m(z, s) = \pi^{-ns}y^{(m+s)} \prod_\nu \Gamma(s + m + q_\nu)$.

If $\mathfrak{q}$ is an integral ideal then the action of the involution $J_{\mathfrak{q}}$ on (4.3) and (4.4) can be computed from the definition (1.20); it is given by the formula

$$K_m^q(s;\mathfrak{a},\mathfrak{b};\eta)|J_{\mathfrak{q}} = (-1)^{mn}\mathcal{N}(\mathfrak{q}\mathfrak{d}^2)^{-s-(m/2)}L_m^q(s;\mathfrak{b},\mathfrak{a}\mathfrak{q}^{-1};\eta). \tag{4.6}$$

We shall need Fourier expansions of the Eisenstein series which can be explicitly written in terms of the Whittaker function $W(y,\alpha,\beta)$. This function is defined by the integral

$$W(y,\alpha,\beta) = \int_0^\infty (u+1)^{\alpha-1}u^{\beta-1}e^{-yu}du$$

which is absolutely convergent for $\mathrm{Re}(\alpha+\beta) > 1$, and

$$W(y,\alpha,-r) = \sum_{i=0}^{r}(-1)^i\binom{r}{i}\frac{\Gamma(\alpha)}{\Gamma(\alpha-i)}y^{r-i} \text{ for } r\in\mathbf{Z}, r\geq 0$$

(compare with the confluent hypergeometric function in 3.2 of chapter 2).

4.2. Proposition. *(On Fourier expansions of the Eisenstein series). Let s be an integer such that $s-q_\nu \leq 0$ for all ν, $\eta(\mathrm{mod}\,\mathfrak{c})$ be a Hecke character of finite order as above (not necessarily primitive) such that $\mathfrak{c}\neq\mathcal{O}$. Then under the above notation there is the following Fourier expansion:*

$$\frac{D_F^{1/2}\mathcal{N}(\tilde{t}_\lambda)\prod_\nu\Gamma(s+m+q_\lambda)}{(-2\pi i)^{n(m+2s)}(-1)^{ns+\{q\}}}L_m^q(z,0;\mathcal{O}_F,\tilde{t}_\lambda^{-1}\mathfrak{d}^{-1};\eta) =$$
$$(4\pi y)^{-q}\sum_{0\ll\xi\in\tilde{t}_\lambda}a_\lambda(\xi,s,y,\eta)e_F(\xi z),$$

where

$$a_\lambda(\xi,s,y,\eta) =$$
$$\sum_{\substack{\xi=\tilde{b}\tilde{c}\\ c\in\mathcal{O}_F, b\in\tilde{t}_\lambda}} \mathrm{sign}\mathcal{N}(\tilde{b})^{m-1}\mathcal{N}(\tilde{b})^{m+2s-1}\eta^*(\tilde{c})\prod_\nu W(4\pi\xi_\nu y_\nu, m+s+q_\nu, s-q_\nu),$$

and $W(y,\alpha,\beta)$ is the Whittaker function.

Proof. Using Fourier transform, one has

$$D_F^{1/2}\mathcal{N}(\tilde{t}_\lambda)\sum_{t\in\tilde{t}_\lambda^{-1}\mathfrak{d}^{-1}}\left(\frac{\overline{z}+t}{z+t}\right)^q\mathcal{N}(z+t)^{-m}|\mathcal{N}(z+t)|^{-2s} =$$
$$\sum_{b\in\tilde{t}_\lambda}\prod_\nu\int_{\mathbf{R}}\frac{\exp(-2\pi i b_\nu t_\nu)\,dt_\nu}{(z_\nu+t_\nu)^{-(m+2q_\nu)}(\overline{z_\nu}+t_\nu)^{-2(s-q_\nu)}},$$

where

$$\int_{\mathbf{R}} \frac{\exp(-2\pi i b_\nu t_\nu)\, dt_\nu}{(z_\nu + t_\nu)^{-(m+2q_\nu)}(\overline{z_\nu} + t_\nu)^{-2(s-q_\nu)}} =$$

$$(-2\pi i)^{m+2s}(-1)^{s+q_\nu} b_\nu^{m+2s-1}\Gamma(s+m+q_\nu)^{-1} W(4\pi b_\nu y_\nu, m+s+q_\nu, s-q_\nu) e_F(bz).$$

Indeed, if we consider the integral

$$f(t) = \int_{-\infty}^{\infty} z^{-\alpha}\overline{z}^{-\beta} \exp(-2\pi i t z)\, dz,$$

which is absolutely convergent for $\mathrm{Re}(\alpha+\beta) > 1$, then application of contour integration shows that

$$f(t) = (2\pi)^{\alpha+\beta} i^{\beta-\alpha}\Gamma(\alpha)^{-1}\Gamma(\beta)^{-1} e^{2\pi t y} \times$$

$$\times \begin{cases} t^{\alpha+\beta-1} e^{-4\pi t y} W(4\pi t y, \alpha, \beta) & \text{if } t > 0 \\ |t|^{\alpha+\beta-1} W(4\pi |t| y, \beta, \alpha) & \text{if } t < 0 \end{cases}$$

(see [Shi 2], pp.84 –85). Therefore

$$D_F^{1/2}\mathcal{N}(\tilde{t}_\lambda) \sum_{t \in \tilde{t}_\lambda^{-1}\mathfrak{d}^{-1}} \left(\frac{\overline{z}+t}{z+t}\right)^q \mathcal{N}(z+t)^{-m}|\mathcal{N}(z+t)|^{-2s} =$$

$$(-2\pi i)^{n(m+2s)}(-1)^{ns+\{q\}} \prod_\nu \Gamma(s+m+q_\nu)^{-1} \times$$

$$\times \sum_{b \in \tilde{t}_\lambda} \mathcal{N}(b)^{m+2s-1} \prod_\nu W(4\pi b_\nu y_\nu, m+s+q_\nu, s-q_\nu) e_F(bz),$$

and the proposition is then easily deduced from the last equality (see the analogous calculation in [Ka3]).

Remark. We shall need only a special case of 4.2, when $q = 0$, $m = k - l$ since the weights k and l are scalars; however, the proposition is applicable for the study of Rankin convolutions of Hilbert modular forms $\mathbf{f}$ and $\mathbf{g}$ of arbitrary integer vector weights $k = (k_1, \cdots, k_n)$ and $l = (l_1, \cdots, l_n)$ satisfying the following condition

$$k_1 \equiv k_2 \equiv \cdots \equiv k_n \bmod 2, \quad l_1 \equiv l_2 \equiv \cdots \equiv l_d \bmod 2.$$

4.3. The integral representation. We set

$$F = \mathbf{f}_0|c(\mathbf{g}) \in \mathcal{S}_k(\mathfrak{c}\mathfrak{m}_0, \psi), \quad G = \mathbf{g}(\chi_{\mathfrak{m}_0})|J_{\mathfrak{c}\mathfrak{m}'} \in \mathcal{S}_l(\mathfrak{c}\mathfrak{m}', \omega^{-1}\chi^{-2}).$$

Then the following *integral representation of Rankin type* holds (see [Shi6], (4.32)):

$$\Psi(s, F, G) = D_F^{1/2}\Gamma(s+1-l)^n \pi^{-ns}\langle F^\rho, V(s-k+1, \chi)\rangle_{\mathfrak{c}\mathfrak{m}'}, \tag{4.7}$$

where

$$V(s, \chi) = G \cdot K_{k-l}^0(s; \mathfrak{c}\mathfrak{m}', \mathcal{O}_F; \psi\omega^{-1}\chi^{-2}).$$

More precisely,

$$\Psi(s,F,G)=D_F^{1/2}\Gamma(s+1-l)^n\pi^{-ns}\sum_{\lambda=1}^{h}\mathcal{N}(\tilde{t}_\lambda)^{s+1-(k+l)/2}\times$$
$$\times\int_{\Gamma_\lambda(\mathfrak{c}\mathfrak{m}')\backslash\mathfrak{H}^n}\overline{F_\lambda^\rho(z)}G_\lambda(z)\,K_{k-l}^0(z,s-k+1;\tilde{t}_\lambda\mathfrak{d},\mathcal{O}_F;\psi\omega^{-1}\chi^{-2})\mathcal{N}(y)^{s-1}\,dx\,dy. \tag{4.8}$$

4.4. Application of the trace operator. There is a method by which the right-hand side of (4.7) can be explicitly calculated. This method is based on the use of the *trace operator* (as in [Man-Pa], [Pa3], [Pa9], [Ar]) and it is described below. Let R be a set of right coset representatives for $W_{\mathfrak{c}\mathfrak{m}'}\backslash W_{\mathfrak{c}\mathfrak{m}_0}$, then for a (not necessarily holomorphic) function $V\in\tilde{\mathcal{M}}_k(\mathfrak{c}\mathfrak{m}',\overline{\psi})$ the trace $V|\mathrm{Tr}_{\mathfrak{c}\mathfrak{m}_0}^{\mathfrak{c}\mathfrak{m}'}\in\tilde{\mathcal{M}}_k(\mathfrak{c}\mathfrak{m}_0,\overline{\psi})$ is defined by the equality: for $x\in G_{\mathbf{A}}, h\in R$

$$(V|\mathrm{Tr}_{\mathfrak{c}\mathfrak{m}_0}^{\mathfrak{c}\mathfrak{m}'})(x)=\sum_{h\in R}\psi(h^{-\iota})V(x\,h). \tag{4.9}$$

Now the inner product in the right-hand side of (4.7) takes the form:

$$\langle F^\rho,V(s-k+1,\chi)\rangle_{\mathfrak{c}\mathfrak{m}'}=\langle F^\rho,V(s-k+1,\chi)|\mathrm{Tr}_{\mathfrak{c}\mathfrak{m}_0}^{\mathfrak{c}\mathfrak{m}'}\rangle_{\mathfrak{c}\mathfrak{m}_0}. \tag{4.10}$$

The trace in the right-hand side can be computed in terms of Fourier expansions using the identity:

$$V|\mathrm{Tr}_{\mathfrak{c}\mathfrak{m}_0}^{\mathfrak{c}\mathfrak{m}'}=\mathcal{N}(\mathfrak{m}'\mathfrak{m}_0^{-1})^{1-(k/2)}V|J_{\mathfrak{c}\mathfrak{m}'}U(\mathfrak{m}'\mathfrak{m}_0^{-1})J_{\mathfrak{c}\mathfrak{m}_0}. \tag{4.11}$$

This identity follows from the explicit description (1.19) of the operator $U(\mathfrak{m}'\mathfrak{m}_0^{-1})$ and the definition (1.20). Indeed, if we use the special choice of a set of representatives for $W_{\mathfrak{c}\mathfrak{m}'}\backslash W_{\mathfrak{c}\mathfrak{m}_0}$, namely,

$$R=\left\{\begin{pmatrix}1&0\\cm_0v&1\end{pmatrix}\ |v\in\mathcal{O}_F/(\mathfrak{m}'\mathfrak{m}_0^{-1})\right\}$$

where cm_0 is an idele such that $\widetilde{cm_0}=\mathfrak{c}\mathfrak{m}_0$, then the desired description follows in view of the definitions of $U(\mathfrak{m}'\mathfrak{m}_0^{-1})$ and $J_\mathfrak{c}$ (see also the analogous formula (2.22) of chapter 2). We transform the integral representation (4.7), taking (4.10) and (4.11) into account, to the following:

$$\Psi(s,F,G)=D_F^{1/2}\Gamma(s+1-l)^n\pi^{-ns}\mathcal{N}(\mathfrak{m}'\mathfrak{m}_0^{-1})^{1-(k/2)}\times$$
$$\times\langle F^\rho,V'(s-k+1,\chi)|U(\mathfrak{m}'\mathfrak{m}_0^{-1})J_{\mathfrak{c}\mathfrak{m}_0}\rangle_{\mathfrak{c}\mathfrak{m}_0}, \tag{4.12}$$

where

$$V'(s,\chi)=V(s,\chi)|J_{\mathfrak{c}\mathfrak{m}'}=(-1)^{nl}\mathbf{g}(\chi_{\mathfrak{m}_0})(K_{k-l}(s;\mathfrak{c}\mathfrak{m}',\mathcal{O}_F;\psi\omega^{-1}\chi^{-2})|J_{\mathfrak{c}\mathfrak{m}_0})=$$

$$(-1)^{nk}\mathbf{g}(\chi_{\mathfrak{m}_0})\mathcal{N}(\mathfrak{c}\mathfrak{m}'\mathfrak{d}^2)^{-s-((k-l)/2)}L_{k-l}(s-k+1;\mathcal{O}_F,\mathcal{O}_F;\psi\omega^{-1}\chi^{-2})$$

according to (4.6).

4.5. Normalized Eisenstein series. We define the normalized Eisenstein series by

$$E_m(s,\eta) = \frac{2^{-n} D_F^{1/2} \Gamma(m+s)^n}{(-4\pi)^{-sn}(-2\pi i)^{n(m+2s)}} L_m^0(s; \mathcal{O}_F, \mathcal{O}_F; \eta). \tag{4.13}$$

Then the Fourier expansion of Proposition 4.2 can be rewritten in the following more covenient form:

$$E_m(s,\eta)_\lambda(z) = \mathcal{N}(\tilde{t}_\lambda)^{-s-(m/2)-1}(4\pi)^n(\mathcal{N}y)^s \sum_{0 \ll \xi \in \tilde{t}_\lambda} a_\lambda(\xi, s, y, \eta) e_F(\xi z),$$

where

$$a_\lambda(\xi, s, y, \eta) =$$
$$\sum_{\substack{\xi = \tilde{b}\tilde{c} \\ c \in \mathcal{O}_F, b \in \tilde{t}_\lambda}} \operatorname{sign}\mathcal{N}(\tilde{b})^{m-1} \mathcal{N}(\tilde{b})^{m+2s-1} \eta^*(\tilde{c}) \prod_\nu W(4\pi \xi_\nu y_\nu, m+s, s)$$

For an integer $s = -r$ such that $0 \le r < m$ we have that

$$a_\lambda(\xi, -r, y, \eta) =$$
$$\sum_{\substack{\xi = \tilde{b}\tilde{c} \\ c \in \mathcal{O}_F, b \in \tilde{t}_\lambda}} \operatorname{sign}\mathcal{N}(\tilde{b})^{m-1} \mathcal{N}(\tilde{b})^{m+2s-1} \eta^*(\tilde{c}) \prod_\nu \sum_{i=0}^r (-1)^i \binom{r}{i} \frac{\Gamma(m-r)}{\Gamma(m-r-i)} (4\pi \xi_\nu y_\nu)^{r-i}$$

in view of the formula in the end of 4.1.

On the other hand, we rewrite the integral representation (4.12) using the normalized Eisenstein series: if we put $m = k - l$ and $\eta = \psi\omega^{-1}\chi^{-2}$ then (4.12) transforms to the following:

$$\begin{aligned} \Psi(s,F,G) = (-1)^{nk} \mathcal{N}(\mathfrak{m}'\mathfrak{m}_0^{-1})^{1-(k/2)} \mathcal{N}(\mathfrak{c}\mathfrak{m}'\mathfrak{d}^2)^{-(s-k+1)-(k-l)/2} 2^n \pi^{n(k-l)} \times \\ \times \langle F^\rho, (\mathbf{g}(\chi_{\mathfrak{m}_0}) E_{k-l}(s-k+1, \psi\omega^{-1}\chi^{-2})) | U(\mathfrak{m}'\mathfrak{m}_0^{-1}) J_{\mathfrak{c}\mathfrak{m}_0} \rangle_{\mathfrak{c}\mathfrak{m}_0}. \end{aligned} \tag{4.14}$$

From this we obtain the following integral representation for the values of the distributions (3.3): for $s = l + r, m = k - l \quad (r = 0, \cdots, k-l-1)$, $\mathfrak{m}_0 \mid \mathfrak{m} \mid \mathfrak{m}'$ we have that

$$\begin{aligned} &\tilde{\mu}_{s,\mathfrak{m}}(\chi) = \\ &\gamma'(\mathfrak{m}') \frac{\langle F^\rho, (\mathbf{g}(\chi_{\mathfrak{m}_0}) E_{k-l}(s-k+1, \psi\omega^{-1}\chi^{-2})) | U(\mathfrak{m}'\mathfrak{m}_0^{-1}) J_{\mathfrak{c}\mathfrak{m}_0} \rangle_{\mathfrak{c}\mathfrak{m}_0}}{\Omega(\mathbf{f})}, \end{aligned} \tag{4.15}$$

where $\Omega(\mathbf{f}) = \langle \mathbf{f}, \mathbf{f} \rangle_{\mathfrak{c}(\mathbf{f})}$, and the constant

$$\gamma(\mathfrak{m}') = 2^{nk} i^{k+1} \alpha(\mathfrak{m}')^{-1} \mathcal{N}(\mathfrak{m}_0)^{(k/2)-1} \mathcal{N}(\mathfrak{c}\mathfrak{d}^2)^{(k+l)/2-1}, \tag{4.16}$$

does not depend on s (but it depends on χ).

In order to prove algebraicity properties and congruences for the values (4.15) we consider for $s = l + r - k + 1$, $0 \le r \le k - l - 1$ the function

$$\Phi(s,\chi) = \mathcal{H}ol\,(\mathbf{g}(\chi_{\mathfrak{m}_0})E_{k-l}(s,\psi\omega^{-1}\chi^{-2})),$$

which is obtained by applying a certain holomorphic projection operator $\mathcal{H}ol$ to the Hilbert automorphic form $\mathbf{g}(\chi_{\mathfrak{m}_0})E_{k-l}(s,\psi\omega^{-1}\chi^{-2})$, which for $s \ne 0$ is not of the holomorphic type (however, it is nearly holomorphic in the sense of Shimura [Shi11] for $r > 0, r \le k - l - 1$). We now recall a description of this operator, which is very similar to that in the one-dimentional case (see [Shi6], [St1], [St2]).

4.6. The holomorphic projection operator. First we describe the vector space, on which this operator is defined. The vector space $\tilde{\mathcal{M}}_k(\mathfrak{c},\psi)$ of C^∞-Hilbert automorphic forms consists of functions $F : G_{\mathbf{A}} \to \mathbf{C}$ satisfying (1.3) – (1.5) of §1 and the following condition, analogous to (1.7): for each $x \in G_{\mathbf{A}}$ with $x_\infty = 1$ there exists a C^∞-function $G_x : \mathfrak{H}^n \to \mathbf{C}$ such that $F(xy) = (G_x|_k y)(i,\cdots,i)$ for all $y = \begin{pmatrix} a & b \\ c & d \end{pmatrix} \in G_\infty$. As in §1, this property allows us to describe C^∞-automorphic forms $F \in \tilde{\mathcal{M}}_k(\mathfrak{c},\psi)$ in terms of C^∞-Hilbert modular forms on $\mathfrak{H}^n$. If we set $F_\lambda = G_{x'_\lambda}$ (for $\lambda = 1,\cdots,h$) then F_λ belongs to the vector space $\tilde{\mathcal{M}}_k(\Gamma_\lambda,\psi)$ of all C^∞-modular forms on $\mathfrak{H}^n$ relative to the congruence subgroup Γ_λ. In other words, the following condition is satisfied:

$$(F_\lambda|_k\gamma)(z) = \psi(\gamma)F_\lambda(z) \text{ for each } \gamma \in \Gamma_\lambda, \text{ and } F_\lambda = \sum_{\xi \in \tilde{t}_\lambda} a_\lambda(\xi,y)e_F(\xi x), \qquad (4.17)$$

where $a_\lambda(\xi,y)$ are C^∞-functions on $(\mathbf{R}_+)^n$.

The map

$$F \mapsto (F_1,\cdots,F_h)$$

defines a vector space isomorphism

$$\tilde{\mathcal{M}}_k(\mathfrak{c},\psi) \xrightarrow{\sim} \oplus_\lambda \tilde{\mathcal{M}}_k(\Gamma_\lambda,\psi).$$

A function $F \in \tilde{\mathcal{M}}_k(\mathfrak{c},\psi)$ will be called a *function of moderate growth* if for all $\lambda = 1,\cdots,h$ and $z \in \mathfrak{H}^n$ and for any $s \in \mathbf{C}$ with sufficiently large $\mathrm{Re}(s) \gg 0$ the integral

$$\int_{\mathfrak{H}^n} F(w)(\overline{w} - z)^{-k-|2s|}\mathrm{Im}(w)^{k+s}d^\times w \qquad (4.18)$$

is absolutely convergent and admits an analytic continuation over s to the point $s = 0$ (compare with (4.3) of chapter 2), where by definition

$$u^{-k-|2s|} = \mathcal{N}u^{-k}|\mathcal{N}u|^{-2s} = \prod_\nu (u_\nu^{-k}|u_\nu|^{-2s}).$$

The following proposition is analogous to Theorem 4.2 of chapter 2.

4.7 Proposition. *Let $F \in \tilde{\mathcal{M}}_k(\mathfrak{c},\psi)$ be a function of moderate growth such that its Fourier expansion $F_\lambda = \sum_{\xi\in\tilde{t}_\lambda} a_\lambda(\xi,y)e_F(\xi x)$ contains only terms with totally positive $\xi \in \tilde{t}_\lambda$. Set for $\xi \gg 0, \xi \in \tilde{t}_\lambda$*

$$a_\lambda(\xi) = \frac{(4\pi)^{n(k-1)}(\mathcal{N}\xi)^{k-1}}{\Gamma(k-1)^n} \int_{(\mathbf{R}_+)^n} a_\lambda(\xi, y) e_F(i\xi y) y^{k-2}\, dy \tag{4.19}$$

and suppose that the integral is absolutely convergent. Also, put

$$\mathcal{H}ol\, F = (\mathcal{H}ol\, F_1, \cdots, \mathcal{H}ol\, F_h), \quad \mathcal{H}ol\, F_\lambda(z) = \sum_{\substack{\xi \in \tilde{\mathfrak{r}}_\lambda \\ \xi \gg 0}} a_\lambda(\xi) e_F(\xi z)$$

Then $\mathcal{H}ol\, F \in \mathcal{M}_k(\mathfrak{c}, \psi)$ and for any cusp form $\mathbf{g} \in \mathcal{S}(\mathfrak{c}, \psi)$ we have that $\langle \mathbf{g}, F\rangle_\mathfrak{c} = \langle \mathbf{g}, \mathcal{H}ol\, F\rangle_\mathfrak{c}$.

Proof of the proposition is carried out in a very similar way to that of the theorem 4.2 of chapter 2 by means of the Poincaré series of two variables (of exponential type) of higher level. Put

$$P_{k,\lambda}(z, w, s) = \sum_{\gamma \in \Gamma_\lambda} \psi(\gamma) J(\gamma, z)^{-k-|2s|}(\gamma(z) + w)^{k-|2s|}, \tag{4.20}$$

where

$$J(\gamma, z) = \det\gamma^{1/2}(cz + d) \text{ for } \gamma \begin{pmatrix} a\, b \\ c\, d \end{pmatrix} \in \Gamma_\lambda,$$

which converges absolutely and uniformly on any compact subset in $\mathfrak{H}^n \times \mathfrak{H}^n$ for $k + \mathrm{Re}(2s) > 2$. These series satisfy the following properties, analogous to (4.9) – (4.10), and (4.12) of chapter 2.

(a) *symmetry*

$$P_{k,\lambda}(z, w, s) = P_{k,\lambda}(w, z, , s) \tag{4.21}$$

(b) *automorphy with respect to both arguments:*

$$P_{k,\lambda}(\gamma(z), \gamma(w), s) = \psi(\gamma_1)\psi(\gamma_2) J(\gamma_1, z)^k J(\gamma_2, w)^k P_{k,\lambda}(z, w, s), \tag{4.22}$$

where $\gamma_i = \begin{pmatrix} a_i & b_i \\ c_i & d_i \end{pmatrix} \in \Gamma_\lambda$, $i = 1, 2$;

(c) *the integral representation:* for all $\mathbf{f} \in \mathcal{S}_k(\mathfrak{c}, \psi)$, $\mathbf{f} = (f_1, \cdots, f_h)$ we have that

$$\langle P_{k,\lambda}(-\overline{z}, w, s), f_\lambda(w)\rangle_{\Gamma_\lambda, w} = c(k, s) f_\lambda(z), \tag{4.23}$$

with

$$c(k, s) = 2^{n(a-s-k)} i^{-nk} \frac{\pi^n}{(k+s-1)^n}.$$

The proof of (4.21) is easily deduced from the symmetry relation

$$J(\gamma, z)(\gamma(z) + w) = J(\tilde{\gamma}, w)(\tilde{\gamma}(w) + z),$$

which is valid for all $\gamma = \left(\begin{smallmatrix} a & b \\ c & d \end{smallmatrix}\right) \in \Gamma_\lambda$ with $\tilde{\gamma} = \left(\begin{smallmatrix} d & b \\ c & a \end{smallmatrix}\right) \in \Gamma_\lambda$, and the above integral formula is carried out similarly to that in Klingen's article [Kl5]. The integration in the left-hand side of (4.23) is reduced by a standard unfolding procedure to that over the whole Hilbert upper half plane

$$\langle P_{k,\lambda}(-\overline{z}, w, s), f_\lambda(w)\rangle_{\mathfrak{c}} = \mathrm{Im} z^s \int_{\mathfrak{H}^n} f_\lambda(w)(\overline{w} - z)^{-k-|2s|}\mathrm{Im}(w)^{k+2s} d^\times w. \qquad (4.24)$$

Then we integrate in (4.24) across each of the variables $w_1, \cdots, w_n$ using the standard integral formula

$$\int_{\mathfrak{H}} f(w)(\overline{w} - z)^{-k-|2s|}\mathrm{Im}(w)^{k+s}\, d^\times w = i^k 2^{2-s-k} \frac{\pi}{k+s-1} f(z),$$

(for one complex variable w).

Now we apply Poincaré series (4.20) and set

$$Q_{k,\lambda}(z, w, s) = c(k, s)^{-1} P_k(-\overline{z}, w, s).$$

We will show that the function

$$\mathcal{H}ol\, F_\lambda(z, s)|_{s=0} = \langle Q_{k,\lambda}(z, w, s), F_\lambda(w)\rangle_{\Gamma_\lambda, w}$$

which is obtained by integration over the variable w for $\mathrm{Re}(s) \gg 0$ and by analytic continuation of the right-hand side to $s = 0$, satisfies all requirements of the proposition, i.e. it coincides with the function $\mathcal{H}ol\, F_\lambda$.

Using the above integral representation we have for $\mathrm{Re}(s) \gg 0$:

$$\langle Q_{k,\lambda}(z, w, s), F_\lambda(w)\rangle_{\mathfrak{c},w} = c(k, s)^{-1}(\mathrm{Im}(s))^s \int_{\mathfrak{H}^n} F_\lambda(w)(\overline{w} - z)^{-k-|2s|}\mathrm{Im}(w)^{k+s}\, d^\times w. \qquad (4.25)$$

Consider the subgroup

$$\Gamma_{\infty,\lambda} = \left\{\gamma = \begin{pmatrix} 1 & b \\ 0 & 1 \end{pmatrix}, b \in \tilde{t}_\lambda^{-1}\mathfrak{d}^{-1} \subset \Gamma_\lambda\right\}$$

then $\mathbf{R}^n/\tilde{t}_\lambda{}^{-1}\mathfrak{d}^{-1} \times \mathbf{R}_+^n$ is a fundamental domain for the action of $\Gamma_{\infty,\lambda}$ on $\mathfrak{H}^n$. Therefore for $\mathrm{Re}(s) \gg 0$ the integral in the right-hand side transforms to the following:

$$c(k, s)^{-1}\mathrm{Im}(s)^{-s} \int_{\mathbf{R}^n/\tilde{t}_\lambda^{-1}\mathfrak{d}^{-1}} \int_{(\mathbf{R}_+)^n} F_\lambda(w) S(\overline{w} - z, \tilde{t}_\lambda^{-1}\mathfrak{d}^{-1}, k+s, s)\, \mathrm{Im} w^{k+s}\, d^\times w, \quad (4.26)$$

where the function

$$S(z, \tilde{t}_\lambda^{-1}\mathfrak{d}^{-1}, k+s, s) = \sum_{b \in \tilde{t}_\lambda^{-1}\mathfrak{d}^{-1}} (z + b)^{-k-|2s|}$$

admits an analytic continuation onto the entire complex plane over s, and has the following Fourier expansion at $s = 0$ (see [Ka3]):

$$S(\overline{w} - z, \tilde{t}_\lambda^{-1}\mathfrak{d}^{-1}, k+s, s)|_{s=0} = \mathcal{N}(\tilde{t}_\lambda)^{-1} D_F^{-1/2} (-2\pi i)^{\{k\}} \prod_{\nu=1}^{n} \sum_{0 \ll \xi \in \tilde{t}_\lambda} \xi^{k-1} e_F(\xi(w - \overline{z})). \tag{4.27}$$

Substituting (4.27) to (4.26) we see that the function $\mathcal{H}ol\, F_\lambda(z,s)|_{s=0}$ transforms to the following

$$\mathcal{H}ol\, F_\lambda(z,s)|_{s=0} = \int_{\mathbf{R}^n/\tilde{t}_\lambda^{-1}\mathfrak{d}^{-1}} \int_{(\mathbf{R}_+)^n} F_\lambda(w) e_F(\xi(z - \overline{w})) \mathrm{Im} w^k d^\times w = e_F(\xi z) \int_{(\mathbf{R}_+)^n} a(\xi, v) e_f(i\xi v) v^k \, d^\times v \quad (w = u + iv, v > 0) \tag{4.28}$$

This equality shows that the the function $\mathcal{H}ol\, F_\lambda(z,s)|_{s=0}$ is holomorphic and satisfies the automorphy conditions with respect to $\Gamma_\lambda(\mathfrak{c})$ since these automorphy conditions are satisfied by the integrand $Q_{k,\lambda}(z,w,s)$ in (4.25). From the definition of $\mathcal{H}ol\, F_\lambda(z,s)|_{s=0}$ we now deduce that

$$\langle g_\lambda(z), \langle Q_{k,\lambda}(z,w,s), F_\lambda(w)\rangle_{\Gamma_\lambda, w}\rangle_{\Gamma_\lambda, z} = \langle\langle Q_{k,\lambda}(z,w,s), g_\lambda(z)\rangle_{\Gamma_\lambda, z}, F_\lambda(w)\rangle_{\Gamma_\lambda, w} = \langle g, F_\lambda\rangle_{\Gamma_\lambda},$$

for any $g \in \mathcal{S}_k(\mathfrak{c}, \psi)$ establishing the proposition 4.7.

§5. Integrality properties and congruences for the distributions.

5.1. In order to prove theorem 2.2 we first establish boundedness of the distributions (3.3) for $s = 0, 1, \cdots, k - l - 1$, and then we prove a certain p-adic congruence involving their values on Dirichlet characters for different s ("generalized Kummer congruences"), which is equivalent to existence of the p-adic analytic function in theorem 2.2. The proof is based on the integral representation (4.15).

5.2. Proposition. *(a) Under the assumptions and notation as in theorem 2.2 for all $r = 0, 1, \cdots, k - l - 1$ the $\mathbf{C}_p$-valued distribution $i_p(\tilde{\mu}_{l+r})$ on the group Gal_S is bounded.*
(b) The following p-adic integrals are equal

$$\int_{\mathrm{Gal}_S} \phi \mathcal{N} x_p^r \, di_p(\tilde{\mu}_l) = (-1)^{rn} \int_{\mathrm{Gal}_S} \phi \, di_p(\tilde{\mu}_{l+r}). \tag{5.1}$$

5.3. To prove this, we consider for $s = l + r - k + 1$ the function

$$\Phi(s, \chi) = \mathcal{H}ol\,(\mathbf{g}(\chi_{\mathfrak{m}_0}) E_{k-l}(s, \psi\omega^{-1}\chi^{-2})), \tag{5.2}$$

which is obtained by applying the holomorphic projection operator $\mathcal{H}ol$ to the Hilbert automorphic form $\mathbf{g}(\chi_{\mathfrak{m}_0}) E_{k-l}(s, \psi\omega^{-1}\chi^{-2})$, which for $s \neq 0$ is not of the holomorphic type. We now use the description of the operator given in Proposition 4.7; the necessary growth conditions are obviously satisfied due to the fact that $\mathbf{g}(\chi_{\mathfrak{m}_0})$ is a cusp form so that the function (5.2) is also a cusp form in $\mathcal{S}_k(\mathfrak{c}\mathfrak{m}', \psi)$. Then the equality (4.15) takes the form

$$\tilde{\mu}_{s,\mathfrak{m}}(\chi) = \frac{\gamma'(\mathfrak{m}')}{\Omega(\mathbf{f})} \langle F^\rho, \Phi(s-k+1,\chi)|U(\mathfrak{m}'\mathfrak{m}_0^{-1})J_{\mathfrak{c}\mathfrak{m}_0}\rangle_{\mathfrak{c}\mathfrak{m}_0}. \tag{5.3}$$

The functions (5.2) apparently satisfy the same compatibility condition as the values $\tilde{\mu}_{s,\mathfrak{m}}(\chi)$ and determine a distribution on Gal_S with values in the spaces of Hilbert automorphic forms of holomorphic type of weight k and character $\overline{\psi}$ (whose level is not fixed, and it increases together with the conductor of χ).

5.4. Fourier coefficients of the function $\Phi(s,\chi)$. Let us consider for $s = l+r-k+1$, $0 \le r \le k-l-1$ Fourier expansions of all functions $\Phi(s,\chi)_\lambda(z)$ (for $\lambda = 1,\cdots,h$) on $\mathfrak{H}^n$ which describe the automorphic form $\Phi(s,\chi)$,

$$\Phi(s,\chi)_\lambda(z) = \sum_{\xi\in\tilde{t}_\lambda} \phi_\lambda(\xi,s,\chi)e_F(\xi z). \tag{5.4}$$

The coefficients $\phi_\lambda(\xi,s,\chi)$ also determine distributions on Gal_S and they can be explicitly computed using Fourier expansions of the normalized Eisenstein series given in 4.5 and the description of the holomorphic projection operator in proposition 4.7.

First, we set

$$\mathbf{g}_{\mathfrak{m}}(\chi)_\lambda(z) = \sum_{0\ll\xi\in\tilde{t}_\lambda} b_\lambda(\xi)\chi^*(\xi\tilde{t}_\lambda^{-1})e_F(\xi z), \quad b_\lambda(\xi) = \mathcal{N}(\tilde{t}_\lambda)^{l/2}C(\xi\tilde{t}_\lambda^{-1},\mathbf{g}) \tag{5.5}$$

(see (1.9)). Also, note that due to proposition 4.2 and to the definition of the normalized Eisenstein series we have for $m = k-l$ and for a non-trivial character $\eta = \psi\omega^{-1}\chi^{-2}$ the following expansion

$$E_m(s,\eta)_\lambda(z) = \mathcal{N}\tilde{t}_\lambda^{-s-(m/2)-1}(4\pi)^{ns}(\mathcal{N}y)^s \sum_{0\ll\xi\in\tilde{t}_\lambda} a_\lambda(\xi,s,y,\eta)e_F(\xi z), \tag{5.6}$$

where

$$a_\lambda(\xi,s,y,\eta) =$$
$$\sum_{\substack{\xi=\tilde{b}\tilde{c}\\ c\in\mathcal{O}_F,b\in\tilde{t}_\lambda}} \mathrm{sign}\mathcal{N}(\tilde{b})^{m-1}\mathcal{N}(\tilde{b})^{m+2s-1}\eta^*(\tilde{c})\prod_\nu W(4\pi\xi_\nu y_\nu, m+s,s).$$

Multiplying the series (5.5) and (5.6) we obtain the Fourier coefficients of the function

$$\mathbf{g}(\chi_{\mathfrak{m}_0})_\lambda(z)E_{k-l}(s,\psi\omega^{-1}\chi^{-2})_\lambda(z) =$$
$$\sum_{\xi=\xi_1+\xi_2} b_\lambda(\xi_1)\chi^*(\xi_1)\mathcal{N}(\tilde{t}_\lambda)^{-s-(m/2)}(4\pi)^{ns}(\mathcal{N}y)^s a_\lambda(\xi_2,y,s,\psi\omega^{-1}\chi^{-2})e_F(\xi z) =$$
$$\mathcal{N}(\tilde{t}_\lambda)^{-s-(m/2)} \sum_{\xi=\xi_1+\xi_2} b_\lambda(\xi_1)\chi^*(\xi_1) \sum_{\substack{\xi_2=\tilde{b}\tilde{c}\\ c\in\mathcal{O}_F,b\in\tilde{t}_\lambda}} \mathrm{sign}\mathcal{N}(\tilde{b})^{m-1}\mathcal{N}(\tilde{b})^{m+2s-1}\times \tag{5.7}$$
$$\times(\psi\omega^{-1}\chi^{-2})(\tilde{c})(4\pi)^{ns}(\mathcal{N}y)^s\prod_\nu W(4\pi\xi_\nu y_\nu, m+s,s)e_F(\xi z).$$

For example, if $r = k - l - 1$, then $s = 0$ and the function $\Phi(0,\chi)$ is itself of holomorphic type and

$$\Phi(0,\chi) = \mathbf{g}_{\mathfrak{m}_0}(\chi)E^0_{k-l}(0,\psi\omega^{-1}\chi^{-2}).$$

We know the Fourier expansion of the holomorphic Eisenstein series $E^0_{k-l}(0,\psi\omega^{-1}\chi^{-2})$ due to Klingen [Kl1], [Kl2].

In the general case the Fourier expansion (5.2) then can be computed using the integral formula of Proposition 4.7 (or by means of the Shimura differential operators (as in [Shi6])). Applying (5.7) for $s = (l+r) - k + 1 = -(k-l-1-r)$, $0 \le r \le k-l-1$ we have that

$$\phi_\lambda(\xi,s,\chi) = \frac{(4\pi)^{n(k-1)}\mathcal{N}\xi^{k-1}}{\Gamma(k-1)^n}\int_{(\mathbf{R}_+)^n}\phi_\lambda(\xi,y,s,\chi)e_F(2i\xi y)\mathcal{N}y^{k-1}\,dy =$$

$$\mathcal{N}(\tilde{t}_\lambda)^{-s-(m/2)}\sum_{\xi=\xi_1+\xi_2} b_\lambda(\xi_1)\chi^*(\xi_1)\sum_{\substack{\tilde{\xi}_2=\tilde{b}\tilde{c}\\ c\in\mathcal{O}_F,\, b\in\tilde{t}_\lambda}} \mathrm{sign}\mathcal{N}(\tilde{b})^{m-1}\mathcal{N}(\tilde{b})^{m+2s-1}\times \tag{5.8}$$

$$\times\,(\psi\omega^{-1}\chi^{-2})(\tilde{c})\prod_\nu P_s(\xi_{2,\nu},\xi_\nu);$$

here $\xi_{2,\nu}$ denotes the image of ξ_2 under the real embedding $\xi \to \xi_\nu$ of F, and $P_s(\xi_{2,\nu},\xi_\nu)$ denotes the polynomial

$$P_s(\xi_{2,\nu},\xi_\nu) = \sum_{i=1}^{-s}(-1)^i\binom{-s}{i}\frac{\Gamma(m+s)}{\Gamma(m+s-i)}\frac{\Gamma(k-1-i)}{\Gamma(k-1)}\xi_{2,\nu}^{-s-i}\xi_\nu^i$$

which has integral coefficients. Let us now substitute $\xi_2 = \xi - \xi_1$; as a result one obtains the following identity: for $r = 0,\cdots,k-l-1$ we have that

$$(\mathcal{N}\tilde{t}_\lambda)^{-(k-l)/2-1+r}\phi_\lambda(\xi,l+r-k+1,\chi) = \mathcal{N}(\xi)u_\lambda(\xi)+$$

$$\sum_{\xi=\xi_1+\xi_2}\chi^*_{\mathfrak{m}_0}(\xi_1\tilde{t}_\lambda^{-1})b_\lambda(\xi_1)\mathcal{N}(\xi_1)^{k-l-1-r}\sum_{\substack{\tilde{\xi}_2=\tilde{b}\tilde{c}\\ c\in\mathcal{O}_F,\, b\in\tilde{t}_\lambda}}(\psi\omega^{-1}\chi^{-2})^*(\tilde{c})\mathcal{N}(\tilde{b})^{1-k+l+2r}; \tag{5.9}$$

here ξ_i are totally positive elements of F $(i = 1,2)$ and the second sum is extended over all decompositions of $\tilde{\xi}_2$ into a product of principial ideals $\tilde{c}\tilde{b}$ with the condition $c \in \mathcal{O}_F, b \in \tilde{t}_\lambda$ and $u_\lambda(\xi)$ is an integral linear combination of the numbers $\chi^*_{\mathfrak{m}_0}(\xi\tilde{t}_\lambda^{-1})b_\lambda(\xi)$ and the values of the Hecke character $(\xi\omega^{-1}\chi^{-2})^*$ (roots of unity).

5.10. Now let us consider the linear functional

$$\mathcal{L} : \Phi \quad\mapsto\quad \frac{\langle F^\rho, \Phi|J_{\mathfrak{c}\mathfrak{m}_0}\rangle_{\mathfrak{c}\mathfrak{m}_0}}{\Omega(\mathbf{f})} \tag{5.10}$$

on the linear space $\mathcal{S}_k(\mathfrak{c}\mathfrak{m}_0,\psi)$. From the Atkin – Lehner theory (in Miyake's form [Miy]) it follows that $\mathcal{L}$ is defined over $\overline{\mathbf{Q}}$, i.e. for a finite number of ideals $\mathfrak{m}_i$ and for fixed algebraic numbers $l(\mathfrak{m}_i) \in \overline{\mathbf{Q}}$ we have that

$$\mathcal{L}(\Phi) = \sum_i C(\mathfrak{m}_i, \Phi) l(\mathfrak{m}_i), \tag{5.11}$$

so that the distributions $\tilde{\mu}_{l+r}$ can be written in the form

$$\tilde{\mu}_{l+r,\mathfrak{m}}(\chi) = \gamma(\mathfrak{m}')\mathcal{L}(\Phi_{r,\mathfrak{m}'}(\chi)), \tag{5.12}$$

where

$$\Phi_{r,\mathfrak{m}'}(\chi) = \Phi(l+r-k+1, \chi)|U(\mathfrak{m}'\mathfrak{m}_0^{-1}) \tag{5.13}$$

and the constant $\gamma(\mathfrak{m}')$ is given by (4.16). Now we see from (5.11) that the properties (a) and (b) of the proposition 5.2 are equivalent to the corresponding statements about the Fourier coefficients $\phi_\lambda(l+r-k+1, \xi, \chi)$ for $\xi \equiv 0 \pmod{\mathfrak{m}'\mathfrak{m}_0^{-1}\tilde{t}_\lambda}$. Indeed, we see that in this case the Fourier coefficients in (5.11) are expressed according to (1.9) in terms of the numbers $\phi_\lambda(l+r-k+1, \xi, \chi)$ as follows:

$$\mathcal{N}(\tilde{t}_\lambda)^{k/2} C(\mathfrak{n}, \Phi(l+r-k+1, \chi)) = \begin{cases} \phi_\lambda(l+r-k+1, \xi, \chi), & \text{if } \xi\tilde{t}_\lambda^{-1} = \mathfrak{n}\mathfrak{m}'\mathfrak{m}_0^{-1} \text{ with} \\ & \text{an integral ideal } \mathfrak{n} \subset \mathcal{O}, \\ 0, & \text{otherwise.} \end{cases} \tag{5.14}$$

Therefore the explicit expression (5.9) implies the following congruences

$$\begin{aligned} \phi_\lambda(l+r-k+1, \xi, \chi) \equiv & (-1)^{rn} \sum_{\xi=\xi_1+\xi_2} \sum_{\substack{\xi_2 = \tilde{b}\tilde{c} \\ c\in\mathcal{O}_F, b\in\tilde{t}_\lambda}} \chi^*(\tilde{c}^{-1}\tilde{b}\tilde{t}_\lambda^{-1})\mathcal{N}(\tilde{c}^{-1}\tilde{b}\tilde{t}_\lambda^{-1})^r \times \\ & \times (\psi\omega^{-1})^*(\tilde{c})\mathcal{N}(\tilde{c})^{k-l-1} \pmod{\mathcal{N}(\mathfrak{m}'\mathfrak{m}_0^{-1})}, \end{aligned} \tag{5.15}$$

for

$$0 \ll \xi_i \in \tilde{t}_\lambda \ (i=1,2), \quad c \in \mathcal{O}_F, \quad b \in \tilde{t}_\lambda, \quad \tilde{\xi}_2 + \mathfrak{m}_0 = \mathcal{O}_F.$$

Now proposition 5.2 and theorem 2.2 are directly deduced from the abstract Kummer congruences (see [Ka3], p.258, or 3.3 of chapter 1); we recall that these congruences give the following criterion for boundedness of $\mathbf{C}_p$-valued distributions.

Let $\{f_i\}$ be a system of continuous functions $f_i \in \mathcal{C}(Y, \mathcal{O}_p)$ in the ring $\mathcal{C}(Y, \mathcal{O}_p)$ of all continuous functions on the compact totally disconnected group $Y = \mathcal{X}_S^\times$ with values in the ring of integers $\mathcal{O}_p$ of $\mathbf{C}_p$ such that the $\mathbf{C}_p$-linear span of $\{f_i\}$ is dense in $\mathcal{C}(Y, \mathbf{C}_p)$. Let also $\{a_i\}$ be a system of elements $a_i \in \mathcal{O}_p$. Then the existence of an $\mathcal{O}_p$-valued measure μ on Y with the property

$$\int_Y f_i \, d\mu = a_i$$

is equivalent to the validity of the following congruences: for an arbitrary choice of elements $b_i \in \mathbf{C}_p$ with only finitely many of the b_i nonzero,

$$\sum_i b_i f_i(y) \in p^n\mathcal{O}_p \text{ for all } y \in Y \ \text{ implies } \sum_i b_i a_i \in p^n\mathcal{O}_p.$$

In our case we take the functions f_i to be all of the functions of the form $\chi \mathcal{N} x_p^s$ with s as in proposition 5.2, and $\chi \in X_S^{\mathrm{tors}}$ Hecke characters of finite order. This system apparently has the dense $\mathbf{C}_p$-linear span. For any finite number of Hecke characters $\chi \in X_S^{\mathrm{tors}}$ we choose sufficiently large ideals $\mathfrak{m}$ and $\mathfrak{m}'$ such that each of these characters is defined modulo $\mathfrak{m}$ and the formula (4.15) for the values of the distributions $\tilde{\mu}_{l+r,\mathfrak{m}}(\chi)$ holds. The proof is then completed using the congruences (5.15) (compare with the analogous argument in §4 of chapter 3).

References

[An1] *Andrianov,A.N.*: Spherical functions for GL_n over local fields and summmation of Hecke series // Mat.Sbornik 83 (1970) 429–451 (in Russian)

[An2] *Andrianov,A.N.*: Euler products attached to Siegel modular forms of degree 2 // Uspekhi Mat. Nauk 29 (1974) 44–109 (in Russian)

[An3] *Andrianov,A.N.*: Symmetric squares of zeta functions of Siegel modular forms of degree 2 // Trudy Steklov Math.Institut USSR. 142 (1976) 22–45 (in Russian)

[An4] *Andrianov,A.N.*: On zeta functions of Rankin type associated with Siegel modular forms // Lecture Notes in Math. N 627 (1977) 325–338

[An5] *Andrianov,A.N.*: On the decomposition of Hecke polynomials for the symplectic group of degree n // Mat. Sbornic 104 (1978) 291–341 (in Russian)

[An6] *Andrianov A.N.*: Euler expansions of the theta transformations of Siegel modular forms of degree n // Mat. Sbornik 104 (1977) 390–427

[An7] *Andrianov,A.N.*: Multiplicative arithmetic of Siegel modular forms // Uspekhi Mat. Nauk 34 (1979) 67–135 (in Russian)

[An8] *Andrianov,A.N.*: Modular descent and the Saito-Kurokawa conjecture // Inv. Math. 53 (1979) 267–280

[An9] *Andrianov,A.N.*: On Siegel modular forms of genus n and weight $n/2$ // J. Fac. Sci. Univ. Tokyo. Ser.IA.28 (1981) 487-503

[An-K] *Andrianov,A.N., Kalinin, V.L.*: On analytic properties of standard zeta functions of Siegel modular forms // Mat. Sbornik 106 (1978) 323–339 (in Russian)

[An-M1] *Andrianov,A.N., Maloletkin, G.N.*: Behaviour of the theta series of degree n under modular substitutions // Izvestija Akad.Nauk. Ser.Matem.39 (1975) 243–258 (in Russian)

[An-M2] *Andrianov,A.N., Maloletkin, G.N.*: Behaviour of the theta series of degree n of indefinite quadratic forms under modular substitutions // Trudy Steklov Mat. Institut USSR 148 (1978) 5–15 (in Russian)

[Ar] *Arnaud B.*: Interpolation p-adique d'un produit de Rankin // CRAS Paris. Ser.1, 299 (1984) 527–530

[A-Le] *Atkin A.,Lehner J.*: Hecke operators on $\Gamma_0(m)$ // Math.Ann. 185 (1970) 134–160

[A-Li] *Atkin A., Li W.-Ch. W*: Twists of newforms and pseudo - eigenvalues of W - opertors // Invent. Math. 48 (1978) 221–243

[Ba] *Baily W.L.*: Eisenstein series on tube domains // In: Problems in analysis. Princeton University Press, 1970

[Bö1] *Böcherer S.*: Über die Fourier – Jacobi – Entwickelung Siegelscher Eisensteinreihen // Math. Z. 183 (1983) 21 – 46

[Bö2] *Böcherer S.*: Über die Fourier – Jacobi – Entwickelung Siegelscher Eisensteinreihen. II. // Math. Z. 189 (1985) 81 – 100

[Bö3] *Böcherer S.*: Über die Fourierkoeffizienten Siegelscher Eisenstenreihen // Manuscripta Math. 45 (1984) 273 – 288

[Bö4] *Böcherer S.*: Über die Funktionalgleichung automorpher L-Funktionen zur Siegelschen Modulgruppe // J. reine angew. Math. 362 (1985) 146–168

[B-Š] *Borevič, S.I., Šafarevič, I.R.*: Number theory // 3d ed. Nauka: Moscow, 1985 (in Russian)

[Ca] *Cassels J.W.S.*: Rational quadratic forms // Acad. Press: London, N.Y., San Francisco, 1978, 413 p.

[Ca-F] *Cassels J.W.S.,Frölich A.(Eds)*: Algebraic number theory // Acad. Press: London, N.Y., San Francisco, 1967, 366 p.

[Co-PeRi] *Coates J., Perrin - Riou B.*: On p-adic L-functions attached to motives over $\mathbf{Q}$ (to appear in Advanced Studies in Pure Math.)

[Co-Schm] *Coates J., Schmidt C.-G.*: Iwasawa theory for the symmmetric square of an elliptic curve // J. reine angew. Math. 375/376 (1987) 104–156

[Co-Schn] *Coates J., Schneider P.*: A note on p-adic distribution, Preprint, Heidelberg, 1985

[De1] *Deligne P.*: Formes modulaires et reprèsentations l-adiques // Sém. Bourb. 1968/69, exp. no 335. Springer-Verlag, Lect. Notes in Math. N 179 (1971) 139–172

[De2] *Deligne P.*: La conjecture de Weil.I // Publ. math. IHES 43 (1974) 273–307

[De-R] *Deligne P.,Ribet, K.A.*: Values of Abelian L-functions at negative integers over totally real fields // Invent. Math. 59 (1980) 227–286

[Ev1] *Evdokimov, S.A.*: Euler products for congruence subgroups of the Siegel modular group of degree 2 // Mat.Sbornik 99 (1976) 483–513 (in Russian)

[Ev2] *Evdokimov, S.A.*: Analytic properties of certain Euler products for congruence subgroups of $\mathrm{Sp}_2(\mathbf{Z})$ // Mat.Sbornik 110 (1979) 369–398 (in Russian)

[Fe] *Feit P.*: Poles and residues of Eisenstein series for symplectic and unitary groups // Memoir AMS 61 (1986) N 346, iv.+89p.

[Fo1] *Fomenko O.M.*: Applications of the theory of modular forms to number theory // In the series “Algebra. Topolology. Geometry.” Vol. 15. VINITI Publ.: Moscow, 1977, 5–91 (in Russian)

[Fo2] *Fomenko O.M.*: On Fourier coefficients of Siegel cusp forms of degree n // Zap. Nauchn. Sem. LOMI 144 (1985) 155–166 (in Russian)

[Fr] *Freitag E.*: Siegelsche Modulfunktionen // Springer-Verlag: Berlin e.a.,1983, x+341p.

[Ga] *Garrett, P.B.*: Pullbacks of Eisenstein series: applications //In: Automorphic forms of several variables. Taniguchi symposium, 1983. Birkhaüser, Boston - Basel - Stuttgart, 1984

[Ge-PSh] *Gelbart S.,Piatetski-Shapiro I.I., Rallis S.*: Explicit constructions of automorphic L - functions // Springer-Verlag, Lect. Notes in Math. N 1254 (1987) 152p.

[G-G-PSh] *Gelfand, I.M., Graev, M.I., Piatgetskij-Shapiro, I.I.*: Representation theory and automorphic functions // In the series: "Generalized functions", Vol. 6. Nauka: Moscow, 1966 (English transl.: W.B.Saunders, Philadelphia, 1969)

[Gr-Z] *Gross B., Zagier, Don B.*: Heegner points and derivatives of L- series // Invent. math. 84 (1986) 225–320

[HSh] *Haran Shai*: p-adic L-functions for modular forms // Compos. Math. 62 (1986) 31–46

[Har1] *Harris M.*: Special values of zeta functions attached to Siegel modular forms // Ann. sci Ecole Norm. Sup. 14 (1981) 77–120

[Har2] *Harris M*: The rationality of holomorphic Eisenstein series series // Invent. Math. 63 (1981) 305–310

[Har3] *Harris M*: Maass operators and Eisenstein series // Math. Ann. 258 (1981) 135–144

[Har4] *Harris M*: Eisenstein series on Shimura varieties // Ann. of Math. 119 (1984) 59 - 94

[He] *Hecke E.*: Theorie der Eisensteinscher Reihen und ihre Anwendung auf Funktionentheorie und Arithmetik // Abh. Math. Sem. Hamburg 5 (1927) 199–224

[Hi1] *Hida H.*: A p-adic measure attached to the zeta functions associated with two elliptic cusp forms. I // Invent. Math. 79 (1985) 159–195

[Hi2] *Hida H*: Galois representations into $\mathrm{GL}_2(\mathbf{Z}_p[[X]])$ attached to ordinary cusp forms // Invent. Math. 85 (1986) 545–613

[H-PSh1] *Howe R., Piatetski-Shapiro I. I.*: A counterexample to the "Generalized Ramanujan conjecture" for (quasi-) split groups // Automorphic Forms, Representations and L- functions. Proc. Symp. Pure Math. AMS. Vol.33. Part 1. Providence: R.I., 1979, 315–322

[H-PSh2] *Howe R., Piatetski-Shapiro I. I.*: Some examples of automorphic forms on Sp_4 //Duke Math. J. 50 (1983) 55–106

[HLK] *Hua Lo- Ken*: Harmonic analysis of functions of several complex variables in the classical domains // Providence: R.I., 1963

[Iw] *Iwasawa K.*: Lectures on p-adic L-functions // Ann. of Math.Studies, N 74. Princeton University Press, 1972

[Ja] *Jacquet H.*: Automorphic forms on GL(2). Part II // Springer-Verlag, Lect. Notes in Math. N 278 (1972)

[Ja-L] *Jacquet H., Langlands R.*: Automorphic Forms on GL(2) // Springer-Verlag, Lect. Notes in Math. N 114 (1972) 548 p.

[K1] *Kalinin V.L.*: Eisenstein series on the symplectic group // Mat. Sbornik 103 (1977) 519–549 (in Russian)

[K2] *Kalinin V.L.*: Analytic properties of the convolution of Siegel modular forms of genus n // Mat. Sbornic 120 (1983) N2, 200 - 206, 286 - 287 (in Russian)

[Ka1] *Katz, N.M.*: p-adic interpolation of real analytic Eisenstein series // Ann. of Math. 104 (1976) 459–571

[Ka2] *Katz, N.M.*: The Eisenstein measure and p-adic interpolation // Amer. J. Math. 99 (1977) 238–311

[Ka3] *Katz, N.M*: p- adic L-functions for CM-fields // Invent. Math. 48 (1978) 199-297

[Ki1] *Kitaoka Y.*: Fourier coeficients of Siegel cusp forms of degree two // Proc. Jap. Acad. A58 (1982) 41–43

[Ki2] *Kitaoka Y.*: Lectures on Siegel modular forms and representations by quadratic forms // Springer - Verlag: Berlin e.a.,1986

[Kl1] *Klingen H.*: Über die Werte der Dedekindschen Zetafunktionen // Math. Ann. 145 (1962) 265–272

[Kl2] *Klingen H.*: Über den arithmetischen Charakter der Fourierkoefficienten von Modulformen // Math. Ann. 147 (1962) 176–188

[Kl3] *Klingen H.*: Zum Darstellungssatz für Siegelsche Modulformen // Math. Z. 102 (1967) 30–43

[Kl4] *Klingen H.*: Über Poincarésche Reihen zur Siegelschen Modulgruppe//Math. Ann. 168 (1967) 157–170

[Kl5] *Klingen H.*: Über Poincarésche Reihen vom Exponentialtyp // Math. Ann. 234 (1978) 145–157

[Ko1] *Koblitz Neal*: p-Adic Numbers, p-Adic Analysis and Zeta Functions // 2nd ed. Springer-Verlag, 1984

[Ko2] *Koblitz Neal*: p-Adic Analysis: A Short Course on Recent Work // London Math. Soc. Lect. Notes Series, N 46. Cambridge University Press: London, Cambridge, 1980

[Ko3] *Koblitz Neal*: p- adic congruences and modular forms of half integral weight // Math. Ann. 274 (1986) 199–220

[Koe] *Koecher M.*: Zur Theorie der Modulformen n-ten Grades. I //Math. Z. 59 (1954) 399–416

[Koj] *Kojima H.*: On construction of Siegel modular forms of degree two // Math. Soc. Japan 34 (1982) 393–412

[Ku] *Kubota T.*: Elementary Theory of Eisenstein Series // Halsted Press: N.Y., 1973

[Ku-Le] *Kubota T., Leopoldt H.-W*: Eine p-adische Theorie der Zetawerte // J- reine angew. math. 214/215 (1964) 328–339

[Kurč1] *Kurčanov P.F.*: Zeta functions of elliptic curves over certain abelial extensions of imaginary quadratic fields // Mat.Sbornik 108 (1977) 56–70 (in Russian)

[Kurč2] *Kurčanov P.F.*: Cohomology of discrete groups and Dirichlet series associated with cusp forms of Jacquet – Langlands // Izvestija Akad. Nauk. Ser. Matem. 42 (1978) 588–601 (in Russian)

[Kurč3] *Kurčanov P.F.*: Local measures connected with cusp forms of Jacquet – Langlands over CM - fields // Mat. Sbornik 108 (1979) 483–503 (in Russian)

[Kur] *Kurokawa N.*: Examples of eigenvalues of Hecke operators on Siegel cusp forms of degree two // Invent. Math. 49 (1978) 149–165

[Kur-Miz] *Kurokawa N., Mizumoto S.*: On Eisenstein series of degree two // Proc. Jap. Acad. A57 (1981) 134–139

[La1] *Lang S.*: Algebraic Number Theory // Addison-Wesley, 1970

[La2] *Lang S.*: Elliptic functions // Addison-Wesley, 1970

[La3] *Lang S.*: Introduction to Modular Forms // Springer-Verlag, 1976

[La4] *Lang S.*: Cyclotomic fields //(Graduate Texts in Math.) Springer-Verlag, N.Y., 1980

[La5] *Lang S.*: Fundamentals of Diophantine Geometry // Springer-Verlag, 1983

[Ll1] *Langlands R.*: On the functional equations satisfied by Eisenstein series // Springer-Verlag, Lect.Notes in Math. N 544 (1976) 337 p.

[Ll2] *Langlands R.*: Problems in the theory of automorphic forms // Springer-Verlag, Lect. Notes in Math. N 170 (1970) 18–61

[Le1] *Leopoldt, H.-W.*: Eine Verallgemeinerung der Bernoullischen Zahlen // Abh. Math. Sem. Univ. Hamburg 22 (1958) 131–140

[Le2] *Leopoldt, H.-W.*: Eine p-adische Theorie der Zetawerte. II // J. reine angew. Math. 274/275 (1975) 224–239

[Li1] *Li W.-Ch.W*: Newforms and functional equations // Math. Annalen 212 (1975) 285–315

[Li2] *Li W.-Ch.W*: L-series of Rankin type and their functional equations // Math. Ann. 244 (1979) 135–166

[Maa] *Maass H.*: Siegel's modular forms and Dirichlet series // Springer-Verlag, Lect. Notes in Math. N 216 (1971)

[Mah] *Mahler K.*: An interpolation series for a continuous function of p-adic variable // J. reine angew. Math 199 (1958) 23–34

[Man1] *Manin Yu.I.*: Cyclotomic fields and modular curves // Uspekhi Mat. Nauk 26 (1971) 7–78 (in Russian)

[Man2] *Manin Yu.I.*: Cusp forms and zeta functions of modular curves // Izvestija Akad.Nauk. Ser.Matem. 36 (1972) 19–66 (in Russian)

[Man3] *Manin Yu.I.*: Explicit formulas for the eigenvalues of Hecke operators // Acta arithm. 24 (1973) 239–249

[Man4] *Manin Yu.I.*: Periods of cusp forms and p-adic Hecke series // Mat. Sbornik 92 (1973) 378–401 (in Russian)

[Man5] *Manin Yu.i.*: The values of p-adic Hecke series at integer points of the critical strip // Mat. Sbornik 93 (1974) 621–626 (in Russian)

[Man6] *Manin Yu.I.*: Non-Archimedean integration and p-adic L-functions of Jacquet – Langlands // Uspekhi Mat. Nauk 31 (1976) 5–54 (in Russian)

[Man6] *Manin Yu.I.*: Modular forms and number theory // Proc. Int. Congr. Math. Helsinki, 1978, 177–186

[Man-Pa] *Manin Yu.I., Panchishkin A.A.*: Convolutions of Hecke series and their values at integral points // Mat. Sbornik, 104 (1977) 617 – 651 (in Russian)

[Maz1] *Mazur B.*: On the special values of L-functions // Invent. Math. 55 (1979) 207–240

[Maz2] *Mazur B.*: Modular curves and arithmetic // Proc. Int. Congr. Math. Warszawa,16–24 August 1982. North Holland: Amsterdam, 1984, 185–211

[Maz-SD] *Mazur B., Swinnerton-Dyer H.P.F.*: Arithmetic of Weil curves // Invent. Math. 25 (1974) 1-61

[Maz-W1] *Mazur B., Wiles A.*: Analogies between function fields and number fields // Am. J. Math. 105 (1983) 507–521

[Maz-W2] *Mazur B., Wiles A.*: Class fields of Abelian extensions of $\mathbf{Q}$ // Invent. math. 76 (1984) 179–330

[Maz-W3] *Mazur B., Wiles A.*: On p- adic analytic families of Galois representations // Compos. Math. 59 (1986) 231–264

[Mi-St] *Milnor J., Stasheff J.*: Characteristic classes // (Ann. of Math. Studies ,N 76) Princeton Univ. Press 1974, 330 p.

[Miy] *Miyake T.*: On automorphic forms on GL_2 and Hecke operators // Ann. of Math. 94 (1971) 174–189

[Miz1] *Mizumoto S.*: Fourier coefficients of generalized Eisenstein series of degree two // Proc. Jap. Acad. A59 (1983) 231–264

[Miz2] *Mizumoto S.*: Fourier coefficients of generalized Eisenstein series of degree two. II // Kodai Math. J. 7 (1984) 86–110

[MF] : Modular functions of one variable. I–VI // Springer-Verlag, Lect. Notes in Math. NN 320, 349, 350 (1973); N 476 (1975); NN 601, 627 (1977)

[Ogg1] *Ogg, A.P.*: Modular forms and Dirichlet series // Benjamin, 1969

[Ogg2] *Ogg, A.P.*: On a convolution of L- series // Invent. Math. 7 (1969) 297–312

[Pa1] *Panchishkin A.A.*: There exist no Ramanujan congruences modulo 691^2 // Mat. Zametki 17 (1975) (in Russian)

[Pa2] –: On Dirichlet series attached to modular forms of integral and half integral weight // Izvestija Akad. Nauk.Ser.Mat.43 (1979) 1145-1157 (in Russian)

[Pa3] –: Symmetric squares of Hecke series and their values at integral points // Mat. Sbornik 108 (1979) 393-417(in Russian)

[Pa4] –: On p-adic Hecke series // In: "Algebra" (Ed.by A. I. Kostrikin), Moscow Univ. Press, 1980, pp. 68–71 (in Russian)

[Pa5] –: Complex valued measures attached to Euler products // Trudy Sem. Petrovskogo 7 (1981) 239–244 (in Russian)

[Pa6] –: Modular forms // In the series "Algebra. Topology. Geometry." Vol. 19. VINITI Publ.: Moscow , 1981, pp.135–180 (in Russian)

[Pa7] –: Local measures attached to Euler products in number fields // In "Algebra" (Ed. by A. I. Kostrikin), Moscow Univ. Press,1982, pp.119–138 (in Russian)

[Pa8] –: Automorphic forms and the functoriality principle // In: "Automorphic forms, representations and L-functions". Mir Publ.: Moscow, 1984, pp. 249–286 (in Russian)

[Pa9] *Pančiškin A.A.*: Le prolongement p-adique analytique de fonctions L de Rankin // CRAS Paris 294 (1982) 51–53; 227–230

[Pa10] *Panchishkin A.A.*: A functional equation of the non-Archimedean Rankin convolution // Duke Math.J.54 (1987) 77-89

[Pa11] *Panchishkin A.A.*: Non-Archimedean convolutions of Hilbert modular forms // Abstracts of the 19th USSR Algebraic Conference, September 1987, Lvov. Vol. 1, 211.

[Pa12] *Panchishkin A.A.*: Non-Archimedean Rankin L-functions and their functional equations // Izvestija Akad.Nauk. Ser.Matem.52 (1988) 336–354(in Russian)

[Pa13] *Panchishkin A.A.*: Convolutions of Hilbert modular forms and their non-Archimedean analogues // Mat. Sbornik 136 (1988) 574–587 (in Russian)

[Pa14] *Panchishkin A.A.*: Non-Archimedean automorphic zeta-functions // Moscow University Press, 1988, 165p. (in Russian)

[PSh] *Piatetski-Shapiro I.I.*: Automorphic Functions and the Geometry of Classical Domains // Phismatgiz: Moscow, 1961 (English transl.: Gordon and Breach, 1969)

[PSh-R] *Piatetski-Shapiro I.I., Rallis S.*: L-functions of automorphic forms on simple classical groups //In: Modul. Forms Symp., Durham, 30 June - 10 July 1983. Chichester, 1984

[Rag1] *Raghavan S.*: Modular forms of degree n and representations by quadratic forms // Ann. of Math. 70 (1959) 446–477

[Rag2] *Raghavan S.*: On an Eisenstein series of degree 3 // J. Indian Math. Soc. 39 (1975) 103–120

[Rag3] *Raghavan S.*: Singular modular forms of degree s //Papers Colloq. S. Ramanujan. – A Tribute. Berlin e.a.,1978, 263–272

[Rag4] *Raghavan S.*: Estimates of coefficients of modular forms and generalized modular relations // Automorphic Forms, Representation Theory and Arithmetic. Papers Colloq., Bombay, 8–15 January 1979. Berlin e.a. 1981, 247–254

[Ran1] *Rankin, R.A.*: Contribution to the theory of Ramanujan's function $\tau(n)$ and similar arithmetical functions. I.II // Proc. Camb. Phil. Soc 35 (1939) 351–372

[Ran2] *Rankin, R.A.*: The scalar product of modular forms // Proc. London math. Soc. 35 (1939) 351–372

[Ran3] *Rankin, R.A.*: The construction of automorphic forms from the derivatives of a given form // J. Ind. Math. Soc. 20 (1956) 103–116

[Res] *Resnikoff H.L.*: Automorphic forms of singular weight are singular forms // Math. Ann. 215 (1975) 173–183

[Res-S] *Resnikoff H.L., Saldaña R.L.*: Some algebraic properties of Fourier coefficients of Eisenstein series of degree two // J.reine angew. Math. (1974) 90–109

[Šaf1] *Šafarevič I.R.*: Zeta Function // Moscow University Press, 1969

[Šaf2] *Šafarevič I.R.*: Basic concepts of algebra // In the series: "Modern Problems of Mathematics. Fundamental Directions", Vol. 11, Moscow: VINITI Publ., 1986

[Sat] *Satake I.*: Theory of spherical functions on reductive algebraic groups over p-adic fields // Publ. Math. IHES N 18 (1963)

[Sch] *Schmidt C.-G.*: The p- adic L-functions attached to Rankin convolutions of modular forms // J. reine angew. Math. 368 (1986) 201–220

[Sel] *Selberg A.*: Bemerkungen über Dirichletsche Reihe, die mit der Theorie der Modulformen nahe verbunden ist // Arch. Math. Naturwid. 43 (1940) 47–50

[Se1] *Serre J.-P.*: Cours d'Arithmétique // Press Univ.France: Paris, 1970; 2nd ed. 1977, 188 p.

[Se2] *Serre J.-P.*: Abelian l-adic representations and elliptic curves // Benjamin Publ.: N.Y. 1968, 195 p.

[Shi1] *Shimura G.*: Introduction to the Arithmetic Theory of Automorphic Functions // Princeton Univ. Press, 1971

[Shi2] *Shimura G.*: On the holomorphy of certain Dirichlet series // Proc. Lond. Math. Soc. 31 (1975) 79–98

[Shi3] *Shimura G.*: The special values of the zeta functions associated with cusp forms // Comm. Pure Appl. Math. 29 (1976) 783–804

[Shi4] *Shimura G.*: On the periods of modular forms// Math. Annalen 229 (1977) 211–221

[Shi5] *Shimura G.*: On certain reciprocity laws for theta functions and modular forms //Acta Math. 141 (1978) 35–71

[Shi6] *Shimura G.*: The special values of zeta functions associated with Hilbert modular forms // Duke Math. J. 45 (1978) 637–679

[Shi7] *Shimura G.*: Arithmetic of differential operators on symmetric domains // Duke Math. J. 48 (1981) 813–843

[Shi8] *Shimura G.*: Confluent hypergeometric functions on tube domains // Math. Ann. 260 (1982) 269–302

[Shi9] *Shimura G.*: Algebraic relations between critical values of zeta functions and inner products // Amer. J. Math. 105 (1983) 253–285

[Shi10] *Shimura G.*: On Eisenstein series // Duke Math. J. 50 (1983) 417–476

[Shi11] *Shimura G.*: Nearly holomorphic functions on hermitian symmetric spaces // Math. Ann. 278 (1987) 1–28

[Sh12] *Shimura G.*: On the critical values of certain Dirichlet series and the periods of automorphic forms // Invent. Math. 94 (1988) 245–305

[Sho] *Shokurov V.V.*: The integrals of Shimura of cusp forms // Izvestija Akad. Nauk. Ser.Mat. 44 (1980) 670–718

[Sie1] *Siegel C.-L.*: Über die analytische Theorie der quadratischen Formen // Ann. Math. 36 (1935) 527–606

[Sie2] *Siegel C.-L.*: Über die Zetafunktionen indefiniter quadratischer Formen // Math. Z. 43 (1938) 682–708

[Sie3] *Siegel C.-L.*: Einführung in die Theorie der Modulfunktionen n-ten Grades // Math. Ann. 116 (1939) 617–657

[Sie4] *Siegel C.-L.*: On the theory of indefinite quadratic forms // Ann. Math. 45 (1944) 577–622

[Sie5] *Siegel C.-L.*: Die Funktionalgleichungen einiger Dirichletscher Reihen //-Math. Z. 63 (1956) 363–373

[Sie6] *Siegel C.-L.*: Berechung von Zetafunktionen an ganzzahligen Stellen// Nachr. Acad. Wiss. Göttingen. II. Math.- Phys. Kl. 10 (1969) 1–16

[Sie7] *Siegel C.-L.*: Über die Fourierschen Koeffizienten von Modulformen // Nachr. Acad. Wiss. Göttingen. II. Math.- Phys. Kl. 3 (1970) 15–56

[St1] *Sturm J.*: Special values of zeta functions and Eisenstein series of half integral weght // Amer. J. Math. 102 (1980) 219–240

[St2] *Sturm J.*: The critical values of zeta-functions associated to the symplectic group // Duke Math. J. 48 (1981) 327–350

[V1] *Višik, M.M.*: Non-Archimedean measures associated with Dirichlet series // Mat.Sbornik 99 (1976) 248-260

[V2] *Višik, M.M.*: Non-Archimedean spectral theory // In the series: "Modern Problems of Mathematics" , Vol. 25, Moscow: VINITI Publ., 1984, 51–114

[Wa] *Washington L.*: Introduction to cyclotomic fields // Springer- Verlag: N.Y. e.a., 1982

[Weil] *Weil A.*: Basic Number Theory // (Die Grundlehren der mathematischen Wissenschaften, Bd.144) Springer-Verlag: Hdlbrg-N.Y.,1973, 312 p.

[We1] *Weissauer R.*: Über die Konstruktion von Spitzenformen mit Hilfe von Thetareihen // Abh. Math. Sem. Hamburg 56 (1986) 133–151

[We2] *Weissauer R.*: Stabile Modulformen und Eisensteinreihen // Springer-Verlag, Lect. Notes in Math. N 1219 (1987)

[We3] *Weissauer R.*: Divisors on the Siegel modular variety // Springer-Verlag, Lect. Notes in Math. N 1240 (1987)

[Za] *Zagier, Don B.*: Modular forms whose Fourier ccoefficients involve zeta-functions of quadratic fields // In: Modular functions.V, Springer-Verlag, Lect. Notes in Math. N 627 (1977) 106–168

Subject Index

analytic function: p-adic analytic function 13
– –: bounded $\mathbf{C}_p$-analytic function 24, 26, 82, 85
Bernoulli distribution 17
Bernoulli number: generalized Bernoulli number 18,20
bounded $\mathbf{C}_p$-analytic function 24, 26, 82, 85
bounded $\mathbf{C}_p$-valued measure 82, 86, 109, 125
bounded growth: function of bounded growth 71
compatibility criterion for distributions 17
complex valued distribution 9, 29, 86, 87, 88
confluent hypergeometric function 60, 64
congruence subgroup 82, 91
convolution of distributions 25
Dirichlet character 2, 3, 24, 51, 53, 82, 83, 86, 88, 112
Dirichlet L-function: normalized Dirichlet L-function 65 ,67 ,69, 107, 112
Dirichlet L-series 2, 112
distribution on a profinite set 16
distribution: normalized distributions 92, 94, 115
distribution: regularized distributions 109, 117
distribution: distribution on Galois group 126
domain of definition of p-adic zeta-functions 2, 23, 81, 117, 125
Eisenstein series: Siegel – Eisenstein series 50, 53, 54, 72, 81, 83, 90
– –: Fourier expansion of the Siegel – Eisenstein series 60, 64, 67
– –: Normalized Siegel – Eisenstein series 66, 67, 94, 100
– –: Klingen – Eisenstein series 58, 72
– –: Eisenstein series in the Hilbert modular case 131
– –: Fourier expansion of the Siegel – Eisenstein series 60, 64, 67
– –: Normalized Eisenstein series in the Hilbert modular case 135, 140
factorization of Hecke polynomials 49
Fourier expansion: positivity property of Fourier expansion 98
– – of Eisenstein series in the Hilbert modular case 131
– – of the Siegel – Eisenstein series 60, 64, 67
Frobenius operator 35, 47, 91
function of moderate growth 71
function of bounded growth 71
Gauss sum 84, 86, 109, 112
Gauss sum of degree m 86
Gauss sum over a totally real field 123
generalized Bernoulli number 18, 20
Hecke algebra: global Hecke algebra 40, 82, 85, 122
– –: eigenfunction of global Hecke algebra 3, 82, 85
– –: local Hecke algebras 87
Hecke character of finite order 6, 116

Hecke couple 45
Hecke operator 46, 85, 121
Hecke polynomial 41
– –: factorization of Hecke polynomials 49
Hecke L-function 116
Hilbert automorphic forms: Hilbert automorphic forms of holomorphic type 6, 117, 121
– – –: Fourier expansion of Hilbert automorphic form 121
Hilbert cusp form: primitive Hilbert cusp form 6, 116, 123
Hilbert modular forms: 6, 117, 120
– – –: Non-Archimedean convolution of Hilbert modular forms 6, 117, 124
- - –: twist operator for Hilbert modular forms 123
– – –: root number for Hilbert modular forms 123, 128
holomorphic projection operator ($\mathcal{H}ol$) 70, 77, 99, 136
inertial degree 10
integral representation of Rankin – Shimura 131, 134
integral representation for the Rankin zeta function 56
interpolation series: theorem on analyticity of interpolation series 14
– –: Mahler's interpolation theorem (Mahler's criterion) 13
Iwasawa algebra 9, 22
Klingen – Eisenstein series 58, 72
Kubota – Leopoldt zeta function 27, 112
Kummer congruences: abstract (generalized) Kummer congruences 9, 19, 113, 116
Maass differentioal operator 63, 77
Mahler's interpolation theorem (Mahler's criterion) 13
Measure: Mazur measure 20, 27, 112, 113
– on a profinite set 18
Mellin transform: non-Archimedean S-adic Mellin transform 9, 22, 24, 81, 82, 125
Modular form: (C^∞-)modular form 70, 77, 90, 100
– –: Siegel modular form 37, 82, 87
– –: Hilbert modular form 6, 117, 120
Newton polygon 15, 16, 24
Non-Archimedean S-adic Mellin transform 9, 22, 24, 81, 82, 125
– analytic functions 25
– convolutions of Hilbert modular forms 6, 117, 124
– standard zeta functions ($\mathcal{D}^{c+}(x,f)$, $\mathcal{D}^{c-}(x,f)$) 81, 85
normalized Siegel – Eisenstein series $G^*(z,s)$ 66, 67, 94, 100
– Dirichlet L-function 65, 67, 69, 107, 112
– Eisenstein series (in the Hilbert modular case) 135, 140
– distributions 92, 94, 115
normalizing factors 67, 70, 93, 115
Petersson scalar product (inner product) 38, 70, 81, 91, 116, 122
Poincaré series 72, 137
positivity property of the Fourier expansion 98
primitive Hilbert cusp form 6, 116, 123

ramification index 9
Rankin zeta function (Rankin type convolution of Siegel modular forms) 50, 87
Rankin – Shimura: integral representation of Rankin – Shimura 131, 134
regularized distributions 109, 117
regularizing factors 109, 110
root numbers (for Hilbert modular forms) 123, 128
Satake q-parameters 5, 40, 42, 49, 82
Siegel – Eisenstein series 54, 81, 83, 90
– – –: normalized Siegel – Eisenstein series ($G^*(z,s)$) 94, 100
Siegel cusp form 38, 81, 83, 86
Siegel modular form 37, 82, 87
– – –: Fourier expansion of Siegel modular form 3, 46, 82
Siegel upper half plane 4, 34
spinor zeta function 6, 42, 50, 85
standard zeta function ($\mathcal{D}(s,f,\chi)$) 3, 4, 42, 43, 81, 82
– – –: analytic properties of the standard zeta function 83, 100
– – –: non-Archimedean standard zeta functions ($\mathcal{D}^{c+}(x,f), \mathcal{D}^{c-}(x,f)$) 81, 85
– – –: normalized standard zeta functions ($\mathcal{D}^*(s,f,\chi), \mathcal{D}^+(s,f,\chi), \mathcal{D}^-(s,f,\chi)$) 3, 81, 83, 86, 92, 100
– – –: special values of the standard zeta functions 83, 85, 92
symplectic group 35
Tate field 2, 9, 12, 23, 81, 125
Teichmüller representative 10
theorem on analyticity of interpolation series 14
–: Weierstrass preparation theorem 15, 24
theta series 50, 88
trace operator 90, 134
totally ramified extension 10
twist operator (for Hilbert modular forms) 123
unramified extension 10
Weierstrass preparation theorem 15, 24
Weyl group 40, 49
Whittaker function 131
C^∞-modular form 70, 77, 90, 100
p-adic zeta function, domain of definition of – – –, 2, 23, 81, 117, 125
p-adic function, continuous – –, 12
p-adic analytic function 13
p-adic power series 13
p-ordinary cusp form 4, 7, 84, 117
S-adic Mellin transform 9, 22, 24, 81, 125
S-adic numbers 11

Vol. 1369: B. Jiang, Ch.-K. Peng, Z. Hou (Eds.), Differential Geometry and Topology. Proceedings, 1986–87. Vl, 366 pages. 1989.

Vol. 1370: G. Carlsson, R.L. Cohen, H.R. Miller, D.C. Ravenel (Eds.), Algebraic Topology. Proceedings, 1986. IX, 456 pages. 1989.

Vol. 1371: S. Glaz, Commutative Coherent Rings. Xl, 347 pages. 1989.

Vol. 1372: J. Azéma, P.A. Meyer, M. Yor (Eds.), Séminaire de Probabilités XXIII. Proceedings. IV, 583 pages. 1989.

Vol. 1373: G. Benkart, J.M. Osborn (Eds.), Lie Algebras. Madison 1987. Proceedings. V, 145 pages. 1989.

Vol. 1374: R.C. Kirby, The Topology of 4-Manifolds. Vl, 108 pages. 1989.

Vol. 1375: K. Kawakubo (Ed.), Transformation Groups. Proceedings, 1987. Vlll, 394 pages, 1989.

Vol. 1376: J. Lindenstrauss, V.D. Milman (Eds.), Geometric Aspects of Functional Analysis. Seminar (GAFA) 1987–88. Vll, 288 pages. 1989.

Vol. 1377: J.F. Pierce, Singularity Theory, Rod Theory, and Symmetry-Breaking Loads. IV, 177 pages. 1989.

Vol. 1378: R.S. Rumely, Capacity Theory on Algebraic Curves. III, 437 pages. 1989.

Vol. 1379: H. Heyer (Ed.), Probability Measures on Groups IX. Proceedings, 1988. Vlll, 437 pages. 1989.

Vol. 1380: H.P. Schlickewei, E. Wirsing (Eds.), Number Theory, Ulm 1987. Proceedings. V, 266 pages. 1989.

Vol. 1381: J.-O. Strömberg, A. Torchinsky, Weighted Hardy Spaces. V, 193 pages. 1989.

Vol. 1382: H. Reiter, Metaplectic Groups and Segal Algebras. Xl, 128 pages. 1989.

Vol. 1383: D.V. Chudnovsky, G.V. Chudnovsky, H. Cohn, M.B. Nathanson (Eds.), Number Theory, New York 1985–88. Seminar. V, 256 pages. 1989.

Vol. 1384: J. Garcia-Cuerva (Ed.), Harmonic Analysis and Partial Differential Equations. Proceedings, 1987. Vll, 213 pages. 1989.

Vol. 1385: A.M. Anile, Y. Choquet-Bruhat (Eds.), Relativistic Fluid Dynamics. Seminar, 1987. V, 308 pages. 1989.

Vol. 1386: A. Bellen, C.W. Gear, E. Russo (Eds.), Numerical Methods for Ordinary Differential Equations. Proceedings, 1987. Vll, 136 pages. 1989.

Vol. 1387: M. Petkovi´c, Iterative Methods for Simultaneous Inclusion of Polynomial Zeros. X, 263 pages. 1989.

Vol. 1388: J. Shinoda, T.A. Slaman, T. Tugué (Eds.), Mathematical Logic and Applications. Proceedings, 1987. V, 223 pages. 1989.

Vol. 1000: Second Edition. H. Hopf, Differential Geometry in the Large. Vll, 184 pages. 1989.

Vol. 1389: E. Ballico, C. Ciliberto (Eds.), Algebraic Curves and Projective Geometry. Proceedings, 1988. V, 288 pages. 1989.

Vol. 1390: G. Da Prato, L. Tubaro (Eds.), Stochastic Partial Differential Equations and Applications II. Proceedings, 1988. Vl, 258 pages. 1989.

Vol. 1391: S. Cambanis, A. Weron (Eds.), Probability Theory on Vector Spaces IV. Proceedings, 1987. Vlll, 424 pages. 1989.

Vol. 1392: R. Silhol, Real Algebraic Surfaces. X, 215 pages. 1989.

Vol. 1393: N. Bouleau, D. Feyel, F. Hirsch, G. Mokobodzki (Eds.), Séminaire de Théorie du Potentiel Paris, No. 9. Proceedings. Vl, 265 pages. 1989.

Vol. 1394: T.L. Gill, W.W. Zachary (Eds.), Nonlinear Semigroups, Partial Differential Equations and Attractors. Proceedings, 1987. IX, 233 pages. 1989.

Vol. 1395: K. Alladi (Ed.), Number Theory, Madras 1987. Proceedings. Vll, 234 pages. 1989.

Vol. 1396: L. Accardi, W. von Waldenfels (Eds.), Quantum Probability and Applications IV. Proceedings, 1987. Vl, 355 pages. 1989.

Vol. 1397: P.R. Turner (Ed.), Numerical Analysis and Parallel Processing. Seminar, 1987. Vl, 264 pages. 1989.

Vol. 1398: A.C. Kim, B.H. Neumann (Eds.), Groups – Korea 1988. Proceedings. V, 189 pages. 1989.

Vol. 1399: W.-P. Barth, H. Lange (Eds.), Arithmetic of Complex Manifolds. Proceedings, 1988. V, 171 pages. 1989.

Vol. 1400: U. Jannsen. Mixed Motives and Algebraic K-Theory. Xlll, 246 pages. 1990.

Vol. 1401: J. Steprans, S. Watson (Eds.), Set Theory and its Applications. Proceedings, 1987. V, 227 pages. 1989.

Vol. 1402: C. Carasso, P. Charrier, B. Hanouzet, J.-L. Joly (Eds.), Nonlinear Hyperbolic Problems. Proceedings, 1988. V, 249 pages. 1989.

Vol. 1403: B. Simeone (Ed.), Combinatorial Optimization. Seminar, 1986. V, 314 pages. 1989.

Vol. 1404: M.-P. Malliavin (Ed.), Séminaire d´AIgèbre Paul Dubreil et Marie-Paul Malliavin. Proceedings, 1987–1988. IV, 410 pages. 1989.

Vol. 1405: S. Dolecki (Ed.), Optimization. Proceedings, 1988. V, 223 pages. 1989. Vol. 1406: L. Jacobsen (Ed.), Analytic Theory of Continued Fractions III. Proceedings, 1988. Vl, 142 pages. 1989.

Vol. 1407: W. Pohlers, Proof Theory. Vl, 213 pages. 1989.

Vol. 1408: W. Lück, Transformation Groups and Algebraic K-Theory. Xll, 443 pages. 1989.

Vol. 1409: E. Hairer, Ch. Lubich, M. Roche. The Numerical Solution of Differential-Algebraic Systems by Runge-Kutta Methods. Vll, 139 pages. 1989.

Vol. 1410: F.J. Carreras, O. Gil-Medrano, A.M. Naveira (Eds.), Differential Geometry. Proceedings, 1988. V, 308 pages. 1989.

Vol. 1411: B. Jiang (Ed.), Topological Fixed Point Theory and Applications. Proceedings. 1988. Vl, 203 pages. 1989.

Vol. 1412: V.V. Kalashnikov, V.M. Zolotarev (Eds.), Stability Problems for Stochastic Models. Proceedings, 1987. X, 380 pages. 1989.

Vol. 1413: S. Wright, Uniqueness of the Injective lll_1 Factor. III, 108 pages. 1989.

Vol. 1414: E. Ramirez de Arellano (Ed.), Algebraic Geometry and Complex Analysis. Proceedings, 1987. Vl, 180 pages. 1989.

Vol. 1415: M. Langevin, M. Waldschmidt (Eds.), Cinquante Ans de Polynômes. Fifty Years of Polynomials. Proceedings, 1988. IX, 235 pages.1990.

Vol. 1416: C. Albert (Ed.), Géométrie Symplectique et Mécanique. Proceedings, 1988. V, 289 pages. 1990.

Vol. 1417: A.J. Sommese, A. Biancofiore, E.L. Livorni (Eds.), Algebraic Geometry. Proceedings, 1988. V, 320 pages. 1990.

Vol. 1418: M. Mimura (Ed.), Homotopy Theory and Related Topics. Proceedings, 1988. V, 241 pages. 1990.

Vol. 1419: P.S. Bullen, P.Y. Lee, J.L. Mawhin, P. Muldowney, W.F. Pfeffer (Eds.), New Integrals. Proceedings, 1988. V, 202 pages. 1990.

Vol. 1420: M. Galbiati, A. Tognoli (Eds.), Real Analytic Geometry. Proceedings, 1988. IV, 366 pages. 1990.

Vol. 1421: H.A. Biagioni, A Nonlinear Theory of Generalized Functions, XII, 214 pages. 1990.

Vol. 1422: V. Villani (Ed.), Complex Geometry and Analysis. Proceedings, 1988. V, 109 pages. 1990.

Vol. 1423: S.O. Kochman, Stable Homotopy Groups of Spheres: A Computer-Assisted Approach. VIII, 330 pages. 1990.

Vol. 1424: F.E. Burstall, J.H. Rawnsley, Twistor Theory for Riemannian Symmetric Spaces. III, 112 pages. 1990.

Vol. 1425: R.A. Piccinini (Ed.), Groups of Self-Equivalences and Related Topics. Proceedings, 1988. V, 214 pages. 1990.

Vol. 1426: J. Azéma, P.A. Meyer, M. Yor (Eds.), Séminaire de Probabilités XXIV, 1988/89. V, 490 pages. 1990.

Vol. 1427: A. Ancona, D. Geman, N. Ikeda, École d'Eté de Probabilités de Saint Flour

XVIII, 1988. Ed.: P.L. Hennequin. VII, 330 pages. 1990.

Vol. 1428: K. Erdmann, Blocks of Tame Representation Type and Related Algebras. XV. 312 pages. 1990.

Vol. 1429: S. Homer, A. Nerode, R.A. Platek, G.E. Sacks, A. Scedrov, Logic and Computer Science. Seminar, 1988. Editor: P. Odifreddi. V, 162 pages. 1990.

Vol. 1430: W. Bruns, A. Simis (Eds.), Commutative Algebra. Proceedings. 1988. V, 160 pages. 1990.

Vol. 1431: J.G. Heywood, K. Masuda, R. Rautmann, V.A. Solonnikov (Eds.), The Navier-Stokes Equations – Theory and Numerical Methods. Proceedings, 1988. VII, 238 pages. 1990.

Vol. 1432: K. Ambos-Spies, G.H. Müller, G.E. Sacks (Eds.), Recursion Theory Week. Proceedings, 1989. VI, 393 pages. 1990.

Vol. 1433: S. Lang, W. Cherry, Topics in Nevanlinna Theory. II, 174 pages.1990.

Vol. 1434: K. Nagasaka, E. Fouvry (Eds.), Analytic Number Theory. Proceedings, 1988. VI, 218 pages. 1990.

Vol. 1435: St. Ruscheweyh, E.B. Saff, L.C. Salinas, R.S. Varga (Eds.), Computational Methods and Function Theory. Proceedings, 1989. VI, 211 pages. 1990.

Vol. 1436: S. Xambó-Descamps (Ed.), Enumerative Geometry. Proceedings, 1987. V, 303 pages. 1990.

Vol. 1437: H. Inassaridze (Ed.), K-theory and Homological Algebra. Seminar, 1987–88. V, 313 pages. 1990.

Vol. 1438: P.G. Lemarié (Ed.) Les Ondelettes en 1989. Seminar. IV, 212 pages. 1990.

Vol. 1439: E. Bujalance, J.J. Etayo, J.M. Gamboa, G. Gromadzki. Automorphism Groups of Compact Bordered Klein Surfaces: A Combinatorial Approach. XIII, 201 pages. 1990.

Vol. 1440: P. Latiolais (Ed.), Topology and Combinatorial Groups Theory. Seminar, 1985–1988. VI, 207 pages. 1990.

Vol. 1441: M. Coornaert, T. Delzant, A. Papadopoulos. Géométrie et théorie des groupes. X, 165 pages. 1990.

Vol. 1442: L. Accardi, M. von Waldenfels (Eds.), Quantum Probability and Applications V. Proceedings, 1988. VI, 413 pages. 1990.

Vol. 1443: K.H. Dovermann, R. Schultz, Equivariant Surgery Theories and Their Periodicity Properties. VI, 227 pages. 1990.

Vol. 1444: H. Korezlioglu, A.S. Ustunel (Eds.), Stochastic Analysis and Related Topics VI. Proceedings, 1988. V, 268 pages. 1990.

Vol. 1445: F. Schulz, Regularity Theory for Quasilinear Elliptic Systems and – Monge Ampère Equations in Two Dimensions. XV, 123 pages. 1990.

Vol. 1446: Methods of Nonconvex Analysis. Seminar, 1989. Editor: A. Cellina. V, 206 pages. 1990.

Vol. 1447: J.-G. Labesse, J. Schwermer (Eds), Cohomology of Arithmetic Groups and Automorphic Forms. Proceedings, 1989. V, 358 pages. 1990.

Vol. 1448: S.K. Jain, S.R. López-Permouth (Eds.), Non-Commutative Ring Theory. Proceedings, 1989. V, 166 pages. 1990.

Vol. 1449: W. Odyniec, G. Lewicki, Minimal Projections in Banach Spaces. VIII, 168 pages. 1990.

Vol. 1450: H. Fujita, T. Ikebe, S.T. Kuroda (Eds.), Functional-Analytic Methods for Partial Differential Equations. Proceedings, 1989. VII, 252 pages. 1990.

Vol. 1451: L. Alvarez-Gaumé, E. Arbarello, C. De Concini, N.J. Hitchin, Global Geometry and Mathematical Physics. Montecatini Terme 1988. Seminar. Editors: M. Francaviglia, F. Gherardelli. IX, 197 pages. 1990.

Vol. 1452: E. Hlawka, R.F. Tichy (Eds.), Number-Theoretic Analysis. Seminar, 1988–89. V, 220 pages. 1990.

Vol. 1453: Yu.G. Borisovich, Yu.E. Gliklikh (Eds.), Global Analysis – Studies and Applications IV. V, 320 pages. 1990.

Vol. 1454: F. Baldassari, S. Bosch, B. Dwork (Eds.), p-adic Analysis. Proceedings, 1989. V, 382 pages. 1990.

Vol. 1455: J.-P. Françoise, R. Roussarie (Eds.), Bifurcations of Planar Vector Fields. Proceedings, 1989. VI, 396 pages. 1990.

Vol. 1456: L.G. Kovács (Ed.), Groups – Canberra 1989. Proceedings. XII, 198 pages. 1990.

Vol. 1457: O. Axelsson, L.Yu. Kolotilina (Eds.), Preconditioned Conjugate Gradient Methods. Proceedings, 1989. V, 196 pages. 1990.

Vol. 1458: R. Schaaf, Global Solution Branches of Two Point Boundary Value Problems. XIX, 141 pages. 1990.

Vol. 1459: D. Tiba, Optimal Control of Nonsmooth Distributed Parameter Systems. VII, 159 pages. 1990.

Vol. 1460: G. Toscani, V. Boffi, S. Rionero (Eds.), Mathematical Aspects of Fluid Plasma Dynamics. Proceedings, 1988. V, 221 pages. 1991.

Vol. 1461: R. Gorenflo, S. Vessella, Abel Integral Equations. VII, 215 pages. 1991.

Vol. 1462: D. Mond, J. Montaldi (Eds.), Singularity Theory and its Applications. Warwick 1989, Part I. VIII, 405 pages. 1991.

Vol. 1463: R. Roberts, I. Stewart (Eds.), Singularity Theory and its Applications. Warwick 1989, Part II. VIII, 322 pages. 1991.

Vol. 1464: D. L. Burkholder, E. Pardoux, A. Sznitman, Ecole d'Eté de Probabilités de Saint- Flour XIX-1989. Editor: P. L. Hennequin. VI, 256 pages. 1991.

Vol. 1465: G. David, Wavelets and Singular Integrals on Curves and Surfaces. X, 107 pages. 1991.

Vol. 1466: W. Banaszczyk, Additive Subgroups of Topological Vector Spaces. VII, 178 pages. 1991.

Vol. 1467: W. M. Schmidt, Diophantine Approximations and Equations. VIII, 217 pages. 1991.

Vol. 1468: J. Noguchi, T. Ohsawa (Eds.), Prospects in Complex Geometry. Proceedings, 1989. VII, 421 pages. 1991.

Vol. 1469: J. Lindenstrauss, V. D. Milman (Eds.), Geometric Aspects of Functional Analysis. Seminar 1989-90. XI, 191 pages. 1991.

Vol. 1470: E. Odell, H. Rosenthal (Eds.), Functional Analysis. Proceedings, 1987-89. VII, 199 pages. 1991.

Vol. 1471: A. A. Panchishkin, Non-Archimedean L-Functions of Siegel and Hilbert Modular Forms. VII, 157 pages. 1991.